Asia-Pacific Environment Monograph 1

STATE, COMMUNITIES AND FORESTS IN CONTEMPORARY BORNEO

Asia-Pacific Environment Monograph 1

STATE, COMMUNITIES AND FORESTS IN CONTEMPORARY BORNEO

Editor: Fadzilah Majid Cooke

ANU
THE AUSTRALIAN NATIONAL UNIVERSITY

E PRESS

ANU
E PRESS

Published by ANU E Press
The Australian National University
Canberra ACT 0200, Australia
Email: anuepress@anu.edu.au
Web: http://epress.anu.edu.au

National Library of Australia
Cataloguing-in-Publication entry

Cooke, Fadzilah M.
State, communities and forests in contemporary Borneo.

ISBN 1 920942 51 3
ISBN 1 920942 52 1 (online)

1. Forest management - Borneo. 2. Forest policy - Borneo.
3. Forests and forestry - Borneo. 4. Forestry and
community - Borneo. 5. Land tenure - Borneo. 6. Land use -
Borneo. I. Title. (Series : Asia-Pacific environmental
monographs).

634.92095983

Cover design by Duncan Beard.
Cover photographs: Lesley Potter and Cristina Eghenter.

Table of Contents

List of Figures   v

List of Tables   v

Foreword   vii

Acknowledgments   ix

Abbreviations   xi

Contributors   xiii

Part I. Introduction

1. Recent Development and Conservation Interventions in Borneo, *Fadzilah Majid Cooke*   3

Part II. Framework and Institutions

2. Expanding State Spaces Using 'Idle' Native Customary Land in Sarawak, *Fadzilah Majid Cooke*   25

3. Native Customary Land: The Trust as a Device for Land Development in Sarawak, *Ramy Bulan*   45

4. Decentralisation, Forests and Estate Crops in Kutai Barat District, East Kalimantan, *Anne Casson*   65

Part III. Local Interventions

5. Community Mapping, Tenurial Rights and Conflict Resolution in Kalimantan, *Ketut Deddy*   89

6. Community Cooperatives, 'Illegal' Logging and Regional Autonomy in the Borderlands of West Kalimantan, *Reed L. Wadley*   111

7. Seeking Spaces for Biodiversity by Improving Tenure Security for Local Communities in Sabah, *Justine Vaz*   133

8. Social, Environmental and Legal Dimensions of Adat as an Instrument of Conservation in East Kalimantan, *Cristina Eghenter*   163

9. The Potential for Coexistence between Shifting Cultivation and Commercial Logging in Sarawak, *Mogens Pedersen and Ole Mertz and Gregers Hummelmose*   181

Part IV. Conclusion

10. Concluding Remarks on the Future of Natural Resource Management in Borneo, *Cristina Eghenter*   197

Index   203

List of Figures

1.1. Map of Borneo with international borders and national divisions 4

2.1. Plantation blocks of the middle and lower Baram, Sarawak 29

4.1. Kutai Region, East Kalimantan 67

4.2. Kutai Barat District, East Kalimatan 68

4.3. Actual and planned oil palm development in Kutai Kartanegara, Kutai Barat and Kutai Timur, March 2000 73

4.4. Location of villages in the PT London Sumatra Plantation Area, East Kalimantan 74

5.1. Community mapping, state mapping and the nature of community 96

5.2. Community mapping activities in Indonesia 98

5.3. Case study locations 102

6.1. Location of the Danau Sentarum National Park 114

6.2. Black pepper prices in Kuching, Sarawak (1992–2000) 119

7.1. Contraction of primary old-growth forest in Sabah's Permanent Forest Estate, 1970–95 135

7.2. Location of Ulu Padas within the Central Bornean Montane Forest ecoregion 136

7.3. Villages and land use classification in the Ulu Padas region 138

7.4. Cultural heritage sites in Ulu Padas State Land 139

9.1. Location of Iban communities and the Sekawi logging camp in the Model Forest-Management Area 184

9.2. Location of cultivated rice fields in the Rumah Chili Area, 1974–79 and 1994–99 186

List of Tables

1.1. Oil palm area in Borneo, 1990–2003 (hectares) 8

Foreword

The name 'Borneo' evokes visions of constantly changing landscapes, but with important island-wide continuities. One of the continuities has been the forests, which have for generations been created and modified by the indigenous population, but over the past three decades have been partially replaced by tree crops, grass or scrub. The loss of forests has been most severe in Sabah, where the plantation model is long established. In Kalimantan, populations have grown and both government-backed and illegal forest clearing have increased exponentially, bringing imminent or more distant threats to traditional livelihoods, but also possibilities to engage with new opportunities. Activities in support of conflict resolution and participatory action research have assumed greater importance and find fertile fields for operation. Before the authoritarian Suharto regime ended in 1998, the role of civil society was quite restricted in Indonesia. Since reformation and democratisation, this has changed, with Indonesia now more liberal than Malaysia. Decentralisation, however, has created its own set of problems. This volume tackles issues of tenure, land use change and resource competition, 'tradition' versus 'modernity', disputes within and between communities, between communities and private firms, communities and government. While there are an equal number of chapters from Kalimantan and East Malaysia, it must be said that there is not equal coverage of the various regions. Three of the four Kalimantan papers are from East Kalimantan, where there is more surviving intact forest than elsewhere.

There are many Borneos: I have my own, as do all researchers on this fascinating island. Crossing the Meratus Mountains in South Kalimantan by motor cycle in 1988, we used old logging roads, the memories of their creeping vines and broken bridges being vividly re-created by Anna Tsing's *Friction* (2005: 29). On the southeast coast I encountered my first oil palm estate with its Sumatran owner, one of the early bridgeheads of that commodity now transforming so much of Borneo. In her introduction to this volume, Majid Cooke has noted that, despite the rapid increase in oil palm planting in Kalimantan, the contributors on the Kalimantan side have not chosen to focus on it. One reason for the lack of discussion is probably that the case studies tend to be located within the hilly borderland of Indonesia and Malaysia, and some are in high mountain areas inherently unsuitable for oil palm, including the sole study set in Sabah. This is the 'Heart of Borneo', especially the large Kayan Mentarang National Park. In Kalimantan, most plantation development lies further south, closer to transport facilities within reasonable distance of the coast. This may be changing, however, with the announcement of a central government-supported 'plantation corridor' along the Indonesia-Malaysia border, in association with road development. A major aim would be to control the illegal logging so graphically described here, but the environmental impacts could be much more serious. The Worldwide Fund for Nature (WWF), a main proponent of the 'Heart of Borneo' conservation initiative, leads the critics of that plan.

Whatever outcomes may still lie in the future, this volume, the first in the series of Asia-Pacific Environmental Monographs, provides much interesting, up-to-date and useful material. I commend it to the reader.

Lesley Potter
The Australian National University
November 2005

Tsing, A.L., 2005. *Friction: An Anthology of Global Connection*. Princeton and Oxford: Princeton University Press.

Acknowledgments

The chapters in this book were originally written for a conference organised by the Resource Management in Asia-Pacific Program at The Australian National University in 2001, under the title 'Resource Tenure, Forest Management and Conflict Resolution: Perspectives from Borneo and New Guinea'. It was made possible with financial assistance from various sources, most notably the Australian Agency for International Development (AusAID). The editor would also like to thank the Earth Island Institute's Borneo Project, whose contribution enabled the attendance of a participant from Sarawak.

All papers have been updated, considerable changes being made by some authors, less by others. They can stand on their own, but contain some common themes. As mentioned in the introductory and concluding chapters, the common themes emphasise the importance of focusing on changes at the local level, of situating local complexities in the larger institutional context, and of the possible gain from such an approach in the search for alternative models of development.

The anonymous readers are to be thanked for their thorough review of the papers and their constructive criticisms. As a result, the introductory chapter was completely rewritten to capture more precisely the similarities and differences between the Indonesian and Malaysian parts of Borneo. Part of the introductory chapter was presented in draft form at the Stockholm Seminars organised by the Swedish non-government organisation Albaeco and the (Swedish) International Foundation for Science at the Royal Swedish Academy of Sciences on 15 June 2005. I benefited enormously from comments made at that seminar.

Special people, however, need mention: Cristina Eghenter, who, apart from being a contributor, took on an additional role as commentator to my introduction and to other chapters in the book; Lesley Potter for her continuing support throughout and for her comments and contribution to the introductory chapter; and last but not least, Colin Filer for his patience in seeing the project through to publication.

Fadzilah Majid Cooke
University of Malaysia Sabah
Kota Kinabalu
October 2005

Abbreviations

CIFOR	Centre for International Forestry Research
DFID	Department for International Development
GIS	Geographic Information System
GPS	Global Positioning System
GTZ	Gesellschaft für Technische Zusammenarbeit
HPH	Hak Pengusahaan Hutan [Logging Concession]
HPHH	Hak Pemungutan Hasil Hutan [Forest Product Harvesting Right]
HTI	Hutan Tanaman Industri [Industrial Timber Plantation]
JVC	Joint Venture Company
KKPKD	Kelompok Kerja Program Kehutanan Daerah [Regional Forestry Program Working Group]
LCDA	Land Consolidation and Development Authority
MFMA	Model Forest Management Area
NGO	Non-Government Organisation
PT	Perseroan Terbatas [Company Limited]
PT KEM	PT Kelian Equatorian Mining
SALCRA	Sarawak Land Consolidation and Rehabilitation Authority
SFI	Sabah Forest Industries
SLDB	Sarawak Land Development Board
USAID	United States Agency for International Development
WWF	Worldwide Fund for Nature

Contributors

Ramy Bulan is Associate Professor and Head of the Centre for Legal Pluralism and Indigenous Law in the Faculty of Law at the University of Malaya in Kuala Lumpur. She works mainly on native title, customary law, and issues relating to minority and indigenous peoples' rights in Malaysia.

Anne Casson is an Associate of the Resource Management in Asia-Pacific Program at The Australian National University. She currently lives in Indonesia and works on a number of forestry sector issues, including illegal logging, forest conversion and oil palm development, forest governance and the process of decentralisation in Indonesia's forest sector.

Fadzilah Majid Cooke is a former Research Fellow of the Resource Management in Asia-Pacific Program at The Australian National University, and now heads the Research Unit for Ethnography and Development in the University of Malaysia Sabah. She has conducted extensive research on the political ecology of forest and coastal resources in Sarawak and Sabah.

Ketut Deddy is the Director of SEKALA, an organisation promoting sound environmental management in Indonesia, and a consultant to various other organisations. His main areas of interest are community mapping, forest governance and forest monitoring.

Cristina Eghenter is an anthropologist who has worked in the Indonesian part of Borneo (Kalimantan) for the last 14 years. She advises WWF Indonesia on issues of community empowerment, sustainable development, collaborative management and equity in conservation, and has particular responsibility for community-related activities in the Kayan Mentarang National Park.

Gregers Hummelmose was formerly with the Institute of Geography at the University of Copenhagen, and is currently employed by the Danish Ministry of Food, Agriculture and Fisheries.

Ole Mertz is coordinator of the Research Network for Environment and Development in the Institute of Geography at the University of Copenhagen. His research interests are traditional farming systems, the management of natural resources by local communities in developing countries, and relationships between research and development assistance within the environmental sector.

Mogens Pedersen was formerly with the Institute of Geography at the University of Copenhagen, and is currently undertaking postgraduate studies in the Department of Geography and International Development at Roskilde University in Denmark.

Justine Vaz is a Ph.D. Candidate in the Department of Geographical and Environmental Studies at the University of Adelaide. She previously worked for WWF Malaysia, where she was active in projects on biodiversity and forest conservation in Sabah.

Reed Wadley is Assistant Professor of Anthropology at the University of Missouri-Columbia. He works on issues of forest resource management, conservation, forest-based agriculture and historical ecology in West Kalimantan.

Part I. Introduction

Chapter One

Recent Development and Conservation Interventions in Borneo[1]

Fadzilah Majid Cooke

Introduction

In the hierarchy of Indonesian and Malaysian official development priorities, Borneo occupies a unique niche. While its peoples and their local political economies are regarded as backward or uncivilised by officials, the natural resources which these same people manage are considered rich.[2] The combination of economic poverty and natural resource wealth provides prime sites for 'development', mostly for the good of the majority or the national good. However, towards the end of the 20th century 'development' changed direction. Through Indonesia's decentralisation policy and Sarawak's land development policy targeted specifically at Native Customary Land, 'development' has been more intensely localised than in earlier decades. One goal of this book is to draw attention to state processes at the end of the 20th and beginning of the 21st centuries that appear to be responding to global economic development in ways that have dramatised the strengths and weaknesses of local political economies and natural resource management. A second objective is to address the changing histories and identities of local communities and institutions as they are reshaped, rejuvenated or weakened in the face of state and economic pressures.

This book evaluates development and conservation interventions that are taking place on the island of Borneo. Its strength lies in its attempt to evaluate change processes affecting both the Indonesian and Malaysian parts of the island (see Figure 1.1). The contributors examine changes associated with two major

---

[1] I wish to express my gratitude to Geoffrey Benjamin and Lesley Potter for their support and ideas for this paper, as well as colleagues at University of Malaysia Sabah, especially Ludwig Kamesheidt of GTZ, for his comments on aspects of scientific forestry, and James Alin for comments on Table 1.1. Nevertheless, responsibility for the final outcome is entirely mine.

[2] Examples abound. For Indonesia, see contributions to Li (1999) concerning myths and assumptions about upland peoples generally (in Borneo and elsewhere in Indonesia) in both conventional and 'green' approaches to development. In the conventional development approach, the creativity, diversity, dynamism and productivity of upland environments are overlooked as programs and policies assume a starting point at or near zero. Some 'green' approaches assume that the subsistence orientation of uplanders is somewhat detached from market production. For Malaysia, various analyses for Sarawak and Sabah, including those by Brosius (1997, 1999), Majid Cooke (1999, 2002), Doolittle (2001, 2004), describe a range of official approaches which suggest their innocence, gullibility and vulnerability due to their 'backwardness'. This in turn justifies increasing state intervention into their physical environments and social lives.

economic activities that have affected Bornean landscapes and livelihoods over the last 30 years; namely, large-scale timber and oil palm production. Reflecting conditions in the field, logging, whether legal or illegal, drew the attention of contributors from Indonesian Borneo in a more fundamental way than oil palm production. By contrast, contributors from Malaysian Borneo took greater heed of changes associated with oil palm than with timber production, without underestimating the continued impact of logging on landscapes and lives. Nevertheless, the chapters resonate with common themes across current topics enabling comparisons on important issues including customary or indigenous tenure, borders and their porosity, the potential for conflict resolution among stakeholders and the role of non-government organisations (NGOs) as intermediaries between 'communities' and the state.

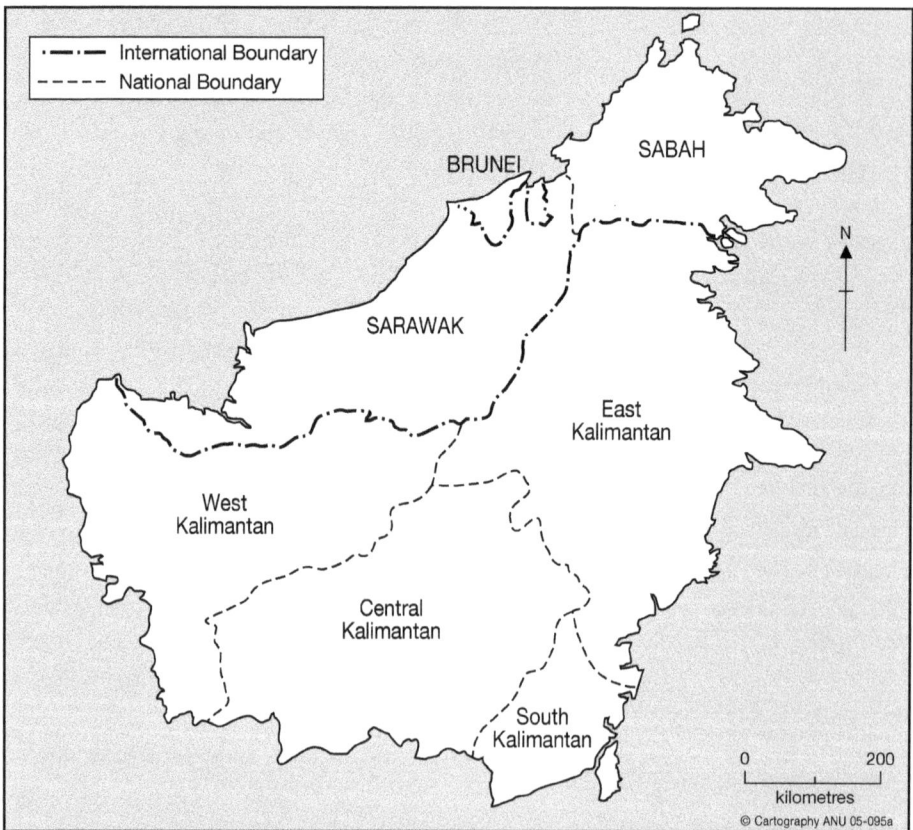

Figure 1.1. Map of Borneo with international borders and national divisions

The authors in this volume have benefited from recent theoretical debates in political ecology, development studies, environmental sociology and social anthropology. Such debates have produced a more critical examination of development, in particular top-down (state-driven) development (Ferguson

1996), of concepts concerning 'community' (Agarwal and Gibson 1999), identity and difference (Li 2003), and of conservation agendas themselves (Brosius 1999). Although benefiting from the philosophical standpoint of post-developmentalism (Rahnema 1992; Escobar 1995), the contributors are consistent in their position of adopting a critical engagement with *alternative development* approaches. This means unpacking notions of 'community', 'participation and empowerment', 'local capacity building and partnership', to name but a few (Friedmann 1992; Brohmann 1996). This engagement is particularly potent since some of the authors are or have been directly involved in implementing these notions on the ground, and therefore experienced the 'unpacking' process directly as they worked in projects supported by NGOs (Deddy, Vaz and Eghenter, Chapters 5, 7 and 8).

From Timber to Oil Palm: State-Driven Development and its Effects on Forests and Customary Land

A decade ago the Bornean territories of Indonesia and Malaysia were described as resource frontier regions (Brookfield et al. 1995). In such regions, 'development' was concerned with economic growth through the export of basic commodities, and the export level had to be maintained even if the commodities changed. In the Bornean context, the export commodities were those that were dependent on the exploitation of natural resources. Timber was produced mainly for export and, once exhausted, was replaced by oil palm. The markets for raw logs and plywood were mainly those of East Asia: Japan, South Korea and Taiwan. In the last decade additional demand for timber came from China and, to a smaller extent, Thailand. Unlike the markets of Europe and, to a lesser extent, the United States, the Asian markets are not pressured into taking account of issues of sustainability, given the relative absence of a strong NGO watchdog movement that the former countries have to contend with.

From the 1970s to the mid-1990s all three regions experienced a period of 'resource boom' (Ross 2001). Booms are characterised by windfall profits during shifts in market price, and because of cheap supply sources (free-standing trees at nominal charges as well as cheap labour) it was possible to capture high 'rents', which represent the margin of profit over and above normal business profit (ibid.). Commodity booms produce a 'get rich quick' mentality among businessmen and a 'boom and bust' psychology among policy makers.

The boom period can be gauged by examining statistics for tropical hardwood production, especially for raw logs. In 1975 Sabah produced 10.1 million cubic metres (mcm) of raw logs, Sarawak only 2.6 mcm and Kalimantan 12.4 mcm. By 1979, which was the peak production year before Kalimantan began its switch to plywood, the four Kalimantan provinces (East Kalimantan being the most important) produced 17.1 mcm, with Sabah and Sarawak producing 9.5 and 7.5 mcm respectively. Sarawak's annual production in the late 1980s averaged 18.8 mcm (ITTO 1990), while raw log production in Sabah stood at 11 and 9.5 mcm in 1988 and 1989 respectively (Chala 2000). In terms of exports of raw logs, Sabah's peak seems to have been in 1977 and 1978 (12.3 and 12.4 mcm respectively). These were also the years of peak production in Kalimantan (13.7 and 14.9 mcm respectively). Sarawak lagged behind and only exceeded Sabah in 1984, with 9.2 mcm produced in that year. By that time Kalimantan no longer exported raw logs, but converted them into plywood. During the 1980s, Indonesia, especially Kalimantan, became the world's leading exporter of tropical plywood (Brookfield and Byron 1990). The boom, which continued into the 1990s, had severe effects on the forest resource base.

There are two indicators that are often used for gauging the environmental viability or otherwise of forestry practices. The first indicator is the volume produced by specific forest patches according to the standards set by a 'sustained yield management' formula, usually expressed as an 'annual allowable cut'. Calculations of annual allowable cut are based either on volume or 'area regulation'. Since reliable growth data are missing in much of Borneo, calculations using either method are educated guesses at best. Moreover, even these guesses are not adhered to. By many accounts, timber production in all three regions consistently exceeded the maximum annual allowable cut many times over (ITTO 1990; Chala 2000; Khan 2001). The second indicator of environmental viability is the condition of the forest after logging. This is a more difficult indicator to work with since it requires knowledge, not only of growth rates, but also of the potential regeneration of species, local soil and weather conditions, and a range of information concerning species tolerance to disturbance at local, landscape and ecosystem levels. Reliable information was largely unavailable or patchy, so management principles were at best only estimates (Majid Cooke 1999; Chala 2000). In Sarawak, Sabah and Kalimantan a 'get rich quick mentality' often meant predatory logging, quick entry without heed to management plans, and complete harvesting of a concession before the licence period expired. The emphasis on speed, together with inadequate data and site surveillance, also affected vulnerable areas such as steep slopes, not normally zoned for logging, and logging beyond the stipulated boundaries was common (Potter 1991; Majid Cooke 1999; Chala 2000; Ross 2001).

A major effect of such predatory logging was forest degradation. Today, more than two thirds of the commercial forest reserves in Sabah consist of degraded logged-over forests, with damage to soil and water quality (as well as loss of fish) so extensive that urgent measures have been required to rehabilitate them (Kollert et al. 2003). Canopy disturbance after logging in Sabah has been estimated at 70 per cent (Chala 2000: 134, citing Nicholson 1979 and Nussbaum 1995). In Kalimantan, logging opened up 80 per cent of the forest canopy (Curran 1999), damaging up to 50 per cent of timber stands in most instances (Tinal and Palenewen 1978: 91, cited in Potter 2005a). Given the ad hoc allocation of licences across the three regions, the question of how much unlogged forest is now left can be difficult to answer. However, an indication can be glimpsed from data on Sabah, where in 1970 it was estimated that there were still 2.7 million hectares unlogged, but by 1996 the unlogged area was only 430 000 hectares (Mannan 1998).

Land conversion for agricultural development, especially oil palm estates, has been an additional factor in forest loss. In three decades approximately one million hectares of forest in Sabah were felled for conversion to oil palm, cocoa and rubber plantations (Chala 2000). In Kalimantan, between 1985 and 1997 approximately 8.5 million hectares were lost, of which half a million hectares were converted to smallholder plantations and 1.7 million to large-scale estates. The balance of 6.3 million hectares was variously accounted for as grassland, scrub and forest regrowth, or as fallow for shifting cultivation (Potter 2005a: Table 4).

Rent seizing through rampant production was made possible by the removal of legal obstacles that could have partially obstructed the process. The centralisation of authority placed the exclusive power to allocate or benefit from logging rights in the hands of individuals or ruling political parties (as in the case of Sabah and Sarawak), or among members of the Suharto family and their cronies, military officials or technocrats (as in Kalimantan) (Peluso 1992; Majid Cooke 1999; Ross 2001). The already limited access rights of indigenous peoples to land acquired through customary claims were further curtailed in 1974 when amendments to the *Land Code* in Sarawak gave individuals or institutions in government the right to extinguish land claimed under customary rights. Similarly, in Indonesia, the *Basic Forestry Law* of 1967, although loosely implemented, was strengthened a decade later through successive regulatory changes which further weakened customary access (Peluso 1992). The legislation gave the central government the authority to grant exploitation rights to private firms directly, bypassing the provincial governments including those of Kalimantan (Peluso 1992; Ross 2001). For Kalimantan, centralisation of the power to allocate concession rights had a major impact on its forests, especially since, in 1982, 67 per cent of Kalimantan's land was classified as either protection/conservation or production forest through the establishment of the *Agreed Forest Land Use Plan* (Ross 1984: 45).

As mentioned earlier, the dynamics of frontier development require that the export economy be maintained. Since good quality logs are now scarce, and surviving timber and plywood mills need to be sustained, part of the supply has to come from elsewhere. Wadley (Chapter 6) refers to the illegal logging taking place at the West Kalimantan/Sarawak border. As well, raw logs from East Kalimantan move into the Sabah town of Tawau (Smith et al. 2003), being allowed into the state as part of an official 'barter trade' (Chala 2000), although the deals associated with the trade may not be officially sanctioned.

Uncertainties as to the future of the timber industry meant that another crop had to be promoted in order to maintain the export industry of the frontier. The crop that filled this need was oil palm. Similar to logging, speed is a characteristic of oil palm development, especially in Sabah and Sarawak. Table 1.1 suggests that areas opened for oil palm increased by an order of magnitude in slightly over a decade.

Table 1.1. Oil palm area in Borneo, 1990–2003 (hectares)

Region	1990	1995	2000	2003	1990–2003 increase (%)
Sabah	276 171	518 133	1 000 777	1 135 100	311.01
Sarawak	54 795	118 783	330 387	464 774	748.20
Kalimantan	87 092	280 247	809 020	1 006 878	1056.10
Total Borneo	418 058	917 163	2 140 184	2 606 752	523.54
Total Indonesia	1 130 000	2 020 000	4 160 000	5 250 000	
Total Malaysia	2 029 464	2 540 087	3 313 393	3 802 040	
Kalimantan/Indonesia %	8	14	19	19	
Sabah/Malaysia %	14	20	30	30	
Sarawak/Malaysia %	3	5	10	12	

Note: figures for Sabah and Sarawak for 2004 were 1 165 412 hectares and 508 309 hectares respectively. Sources: Casson 1999, Figure 1; Direktorat Jenderal Perkebunan 1990–2004; Kalimantan Barat 2003; Kalimantan Selatan 2003; Kalimantan Tengah 2003; Kalimantan Timur 2003; Malaysian Palm Oil Board 2004; Lesley Potter, personal communication.

Table 1.1 suggests that between 1990 and 2003, Borneo became an important area for oil palm in both Malaysia and Indonesia. For Malaysia, 30 per cent of the total land covered by oil palm in 2003 was located in Sabah, followed by 12 per cent in Sarawak. For Indonesia, Kalimantan is not as important as Sumatra, but the expansion in terms of area opened to oil palm has been phenomenal. In 2003, of the 5.25 million hectares of land under oil palm in Indonesia, approximately 19 per cent was located in Kalimantan compared to 72 per cent in Sumatra. However, amongst the three regions of Borneo, the biggest leap was made by Kalimantan with a 1056 per cent increase between 1990 and 2003. Although there was a certain amount of hesitancy in oil palm investment in Kalimantan up until 1998 (Casson 1999), the pace has recently quickened. However, there are still a large number of companies who are only interested in removing the timber from lands granted to them, rather than in planting crops (Potter 2005b).

Sabah and Sarawak also registered great expansion during the same period. Sabah's oil palm area increased by approximately 311 per cent, and Sarawak by approximately 748 per cent. Sarawak has ambitions of doubling its area under oil palm to 1 million hectares by the year 2010 (Deputy Chief Minister of Sarawak cited in *Daily Express*, 14 May 2005; Majid Cooke, Chapter 2). In sum, the total area planted to oil palm in Borneo for 2003 was 2.61 million hectares — 43 per cent in Sabah, 39 per cent in Kalimantan, and 18 per cent in Sarawak.

Although palm oil prices fluctuate, periods of high prices more than compensate for the bad times, so that the crop is regarded as 'green gold' by many (*Daily Express,* 23 and 26 February 2005). One explanation for the push for rapid expansion is that yield per hectare of oil palm is not increasing, and may in fact be declining (Thomas Mielke cited in *Daily Express*, 8 April 2005),[3] so that in order to make maximum profit, expanded hectarage is necessary.

If centres of decision making in the frontier regions of Indonesia and Malaysia, and in the capital cities of Jakarta or Kuala Lumpur, have acted against meaningful conservation in the 1990s (Curran 1999), they are similarly positioned in the 21st century. In Sabah, so entrenched is oil palm in the development equation that a local opposition political party promised to open up more land for oil palm in order to win votes (*Daily Express*, 9 February 2004).[4] In Sarawak, legal and administrative changes under Konsep Baru (New Concept)[5] removed earlier obstacles to converting land claimed under customary access rights into oil palm plantation blocks (Majid Cooke, Chapter 2). In all three regions oil palm is regarded by many small farmers (though not NGOs) as an economic saviour, rather than the environmental vandal portrayed by international conservationists. Intensified localised development pushes the indigenous land movement throughout Borneo in novel directions. Under these circumstances, steering an alternative path has not been, and will not be, easy.

Conservation and the Search for Alternatives

The search for alternatives has engaged diverging interests and philosophies in many unruly alliances and practices, and differences are often papered over. Conservationists who regard conservation goals as non-negotiable tend to view poverty alleviation as a means to an end. From this perspective, the main objective is conservation, so that a major task is to work out the most efficient strategies to achieve this end. This has led to many income-generating projects, such as promoting the 'extraction' of non-timber forest products, and allowing for 'traditional use' zones in national parks. Poverty alleviation as a route to conservation is not a satisfactory position for those who regard development as a right in itself, and this perspective is often coupled with the notion that conservation is a 'neo-colonial' project aimed at keeping the South poor (Fisher 2000). The problem with the latter view is that 'development' here refers to

[3] Thomas Mielke is the Director of an independent research organisation that publishes *Oil World*.
[4] Because of the political and economic factors that work against parties that do not belong to the ruling coalition in Malaysia generally and Sabah specifically, the *Bersekutu* party did not win in the 2004 State elections. Although the party did not get elected, the idea of converting forest reserves to oil palm may be based on the perception that such a strategy will win votes, not lose them – something politicians would be wary about if there were a larger conservation-conscious electorate than is presently in existence in Sabah.
[5] 'New Concept' is a term used to capture the myths around modernity and the efficiency of large-scale, commercial enterprises, especially plantation agriculture in a joint venture program involving private corporations and native peoples (see Majid Cooke 2002; also Bulan, Chapter 3).

economic development, and provides a convenient platform for supporting the suppression of other kinds of development, especially political or social development. In the hierarchy of development priorities in Indonesia and Malaysia, community development in political and social terms is at best pushed to the background, at worst something to be controlled, manipulated or watched over. The need to control political and social development finds affinity among some donor agencies — especially those who, for decades, have viewed development in terms of 'techniques' of economic management, such as export promotion, debt-service management, control of public spending, and market liberalisation. Donor emphasis on techniques was supported by a 'dominant mindset that gives little consideration to socio-cultural issues' (Nelson 1995: 162, 171), but with the discovery of 'civil society' this mindset might have changed, albeit in directions amenable to bureaucratic routinisation (Nelson 1995; Howell and Pearce 2001). The notion of participatory development is a good example for drawing attention to this process. Participatory development originated as a critique of the 'top down' development approach, which had already become routinised, especially in large multilateral aid agencies, but still tended to emphasise community participation in the implementation, rather than the design of projects, instead of formulating alternative development approaches. Smaller bilateral aid agencies may experience fewer constraints and have more options for change (Brohman 1996).

For some NGOs who view past donor efforts at promoting 'development' as a factor contributing to environmental degradation, new alliances with donors represent a pragmatic way of making an entry into the policy debate, despite suspicions of 'neo-colonialism'. Touting the platform of 'neo-colonialism' is an easy way for some elements of the state and society to harness nationalist sentiments against activities that may lead to a questioning of existing power relations and dominant ways of 'doing development'. In many instances, conservation efforts that emphasise 'empowerment' or 'participation' are not as effective in questioning existing power relations as the neo-colonial rhetoric. Equating conservation with 'neo-colonialism' is intrinsic among some NGOs of the South, so that in some instances conservation NGOs are torn between advancing the agenda of global equality and that of conservation (Khor 1993; Shiva 1993). A concern with the former often finds NGOs forming alliances with governments of whom they had previously been critical. However, such alliances are often uneasy ones, confined to specific issues and therefore sporadic. This is because, while some Southern NGOs are annoyed over what they regard as Northern NGOs' insensitivity towards the historical roots of global inequality and environmental injustice, their own governments use 'neo-colonialism' to ward off international criticism regarding their suppression of political and social development. Reminding the nation of 'neo-colonialism' enables many regimes to pursue 'development' as usual.

Under such conditions, a marriage between international conservation NGOs and donors may be a strategic move for entry into a recipient country, but the marriage is also fraught with the danger of 'co-optation' through routinisation (Howell and Pearce 2001: 94–7). For some national NGOs, cooperating with donor agencies is a way of taking the agenda of conservation (which in the South is inseparable from issues of social justice) out of restrictive state control into the international arena (Nelson 1995). In this case, the environment becomes a safe mechanism for advancing issues of citizenship and a mildly disguised critique concerning government accountability and transparency. In general, NGOs currently have more freedom and a higher status in Indonesia than in Malaysia, where the governments tend to be very suspicious of them (Eldridge 1996; Majid Cooke 2003a; Weiss 2003). However, this role for NGOs in Indonesia is very new — during the Suharto regime they were quite restricted in their activities.

In situations where states are not sympathetic to conservation, then creating strategic alliances with state institutions and donors adds additional and much needed clout to NGOs. Casson (Chapter 4) writes about options available to new administrative districts (created as a result of Indonesia's regional autonomy law) in pursuing 'development' objectives (strictly economic) and in preserving 'old ways'. Acquiring relative independence means new responsibilities: the district treasury has to be filled, poor infrastructure upgraded, and long-term planning for sustainable development put in place. Casson's chapter is interesting in a number of ways. It describes a Kalimantan local district government's attempt at more responsible management of resources, and the ways stakeholders — including government officials, community and private sector representatives, *adat* or customary leaders, as well as donor agencies and NGOs — align themselves with one another. The chapter takes issue with the common-sense view about the enhanced potential for conservation in a more decentralised system of decision making. At least in the initial years of autonomy, the district of Kutai Barat showed little evidence of being more environmentally responsible in its development plans than when administration was more centralised.

Exposed to only one form of development, many rural communities have internalised development in economic terms, negotiating top-down and unequal power relations, producing effects that may be detrimental to social and political development, albeit not always of their own choosing (Li 2001; Majid Cooke 2002). In such a scenario, social and political development appear relatively unimportant, and conservation then finds a difficult terrain. Wadley's 'borderlanders' (Chapter 6) are expert negotiators of state boundaries dividing Sarawak from West Kalimantan, which are made porous through kin, labour and commercial networks. As a result, state borders may not carry the nationalistic meaning they are supposed to have. Because of such networks, 'illegal logging' by community cooperatives takes on a different meaning.

Working with Malaysian logging concessionaires (*tukei*) is not an issue for these cooperatives because of their intermediary position as borderlanders.

At the community level, another way of dealing with unequal power relations is to engage in activities which, to outsiders, may appear detrimental to the communities' own survival in the long term. Vaz's work (Chapter 7) unravels entrenched views about 'harmonious' communities whose identities are inextricably linked to their environment. The weakening of Lundayeh control of land held under Native Title at Long Pasia produced a divided community. Some groups resorted to ways of earning a living that were tantamount to rendering their land vulnerable to exploitation by outsiders. They did so by providing outsiders with access to local hunting areas or fishing spots and by allowing destructive methods to be used. The once open borders between Kalimantan and Sabah, which allowed Lundayeh families to maintain their kinship links, became less porous as relatives returning from Kalimantan and elsewhere were no longer accorded access to ancestral land.

On Being Indigenous

The word 'indigenous' has been deflated of meaning when political leaders in both Indonesia and Malaysia claim that, given the multitudes of ethnicities and identities in the two countries, 'everyone is indigenous'.[6] Denying ethnic difference may be useful in processes of mobilising allegiance to or support for political centres, and of underplaying localistic attachments to place. As an analytical category, being indigenous is linked with several characteristics, two of which are important for our purpose. Indigenous cultural traditions are associated with an attachment to place most likely derived from not having a migratory family history. Such traditions are further characterised as an attitude of not being able to objectify place, or of not being able to regard place as a commodity (Benjamin 2002, 2005). It is this attachment to place, and a relative inability to treat home places as an exploitable commodity, that accounts for individuals' lack of economic competitiveness, while those who are culturally exogenous, because of a migratory family history, are able to objectify and thus exploit place. For example, the popular marketing of ecological tourism depends on an individual's capacity to regard her/his surrounds as a commodity, implying the build-up of cultural or emotional distance between individuals and their environment, sufficient for them to regard their environment as a marketable product. However, an attachment to place cannot be taken for granted; it exists

[6] There is an interesting point being made about being indigenous that has to do with cultural content (see Benjamin 2005). Indigenous attitudes and orientations, according to Benjamin, 'are coded in the habitus of daily life, and transmitted tacitly rather than by formal teaching...embedded in patterns of language-use, kinship, religious action, customary clothing, music, or vernacular architecture. Inheritance, not innovation, is the mode of cultural communication in such cases, and the *gemeinschaftlich* will far outweigh the *gesellschaftlich* as the locus of authority' (Benjamin 2005: 7). However, what the political leaders are engaging in here is what Benjamin refers to as 'indigenism'.

in varying degrees of intensity among individuals. Paradoxically, in the search for alternatives, the attachment to place has been a useful starting point for those involved in the indigenous land-rights movement, as it has been for the environmental movement.

An attachment to place may, however, be assumed to be present among most groups, and attention can then be focused on their claims for access to land on the grounds of being 'indigenous'. The indigenous land-rights movement argues that one characteristic that binds most local groups who claim indigenous status is insecurity of tenure. Insecurity of tenure is the common experience of all groups dealt with in this volume. Land managed under customary access may be recognised in the various land laws of the two countries: Native Customary Land in Sarawak, Native Title Land in Sabah or land managed under *adat* in Kalimantan, because of ideologies of legal pluralism from the colonial era. But such recognition was only a minor concession to the larger process of converting all 'unoccupied land' to state land (Peluso and Vandergeest 2001). This conversion of all 'unoccupied land' (which may in fact be land left fallow) to state land is referred to as a 'fundamental error' by Majid Cooke in Chapter 2. In recent times, this 'error' has been systematically contested in the courts in Sarawak (Majid Cooke 2003b), as well as in Peninsular Malaysia.[7] The idea is to confine access rules which are embedded in place to a few regulations that the State recognises or understands, with a view to achieving simplified titling in the long term (see Scott 1998). In all three regions of Borneo, land claimed under customary use is secure only when it is titled. In this volume, Majid Cooke, Bulan, Eghenter, Deddy and Vaz all refer to the insecurity of tenure of native land. In view of the elephantine nature of the land administration machinery, the process of titling may take decades in some cases. In the meantime, customary land remains state land and subject to 'development' at the discretion of the state.

In post-colonial times, recognition of customary access is made more restrictive, with rights and entitlements being decided upon by the state; such access becomes visible only so that lands can be targeted for 'development' at the discretion of the state, as under the Konsep Baru in Sarawak (Majid Cooke 2002). In Indonesia, it is only when indigenous groups qualify as enduring *adat* communities, engaging in prescribed 'traditional' livelihood and management practices, that they are recognised as rightful and responsible managers of their land (Li 2003; Eghenter, Chapter 8). The quest for secure title among local communities is therefore understandable.

---

[7] In Peninsular Malaysia, on 14 September 2005, the Orang Asli (Temuan) won what was regarded as a landmark case at the Appeals Court against their being evicted to make way for the Kuala Lumpur Airport highway development. Referred to as the Sagong Tasi case, the Court recognised that the Temuan owned their land through customary title and ordered compensation be paid to them for being treated 'in a most shoddy, cruel and oppressive manner' (Judgement of Gopal Sri Ram, Justice of the Court of Appeal, Malaysia, 14 June 2005 — *Rayuan Sivil* No. B–02419 2002; *Daily Express*, 21 September 2005).

From a local perspective, there is an aspect of conservation interest that looks towards local groups as potential providers of alternative models for living with (as opposed to wanting to control) nature, and this can be useful for advancing claims to secure titles. The combination of local and universal interest in 'place' has been translated into a range of practices, which include research into indigenous or traditional ecological knowledge or local management systems (Berkes and Fowles 1998; Ellen et al. 2000), and participatory resource management, of which community mapping is an important part. Conservation then becomes the umbrella for a diverse range of interests.

Many lessons have been learnt about community desires for conservation. The desire is, first and foremost, fuelled by the potential for generating income (see Filer 1997). Interest in conservation comes only after the relationship between income and sustainable use becomes apparent. Benefits to stakeholders (not necessarily cash) are important incentives for conservation to be successful. However, benefits such as participation and community empowerment, regarded as important for strengthening community capacity to uphold a sustainable society in general, and sustainable resource use in particular, have proved elusive. Among the intended beneficiaries of these approaches, many may remain unconvinced. In many 'participatory' projects participation is encouraged at the implementation level, while planning remains the prerogative of an educated elite. Participation, in these settings, can usefully be viewed as a rhetorical tool designed to influence the *environment in which decision makers act*, rather than changing the decision-making environment at the local level (MacLean 2000). Even in terms of material benefits, examples from India of 'joint forest management' projects suggest that income from non-timber forest products, normally valuable to villagers, now has to be shared with the Forest Department, while villagers do all the work (Sarin 1999; Fisher 2000). There is evidence as well that a large part of the benefit from such projects is being captured by better-off people within these communities.

There are additional risks to participation. Wadley (Chapter 6) has found that providing local communities with the opportunity to decide and control their own development, as has happened since the fall of Suharto's New Order Regime, may not necessarily lead to an increase in local concerns for conservation in West Kalimantan. In fact, decentralisation has resulted in a simultaneous increase in official corruption.

Specific risks can be gleaned from initiatives in community mapping. Participation through community mapping is risky and rarely a straightforward business, as Deddy observes in Chapter 5. First there is the tension between the process of map making and the need to produce maps as a product. Is it the process or the map that is important? An emphasis on map making does not contribute to community participation. The process of map making, if participatory, should involve all members of the community, regardless of social status, age or gender. This type of participation may touch on issues of power/gender relations, and may have implications for the unresolved question of access rights among return migrants. Community mapping may bring to the surface contested claims and titles, the potential being created for the exclusion or inclusion of claims and entitlement, as Wadley found in West Kalimantan. Second, contradicting their original intention, community maps are at risk of being used for exclusionary purposes; for example, to support the interests of powerful community members in promoting destructive logging or plantation development. When maps are used for exclusionary purposes, issues of community access may be ignored, as are claims of less powerful groups against their neighbours.

Ideas about participation also underlie interest in local management systems. The subtext of the interest in local management systems is to draw attention to potential alternatives to the top-down approaches in natural resource management. Eghenter (Chapter 8) discusses the advantages of taking local management systems seriously and respecting institutional (*adat*) capacity for managing local access to resources. Where local institutions are changing or weakening from market or other forms of largely external pressure, then she recommends strengthening them. Only when local capacities are developed will participation be real. At the Kayan Mentarang National Park mechanisms have been put in place for an inter-*adat* institutional coordination body of elected members of different customary councils to actively manage the conservation area. Under this arrangement, the central and regional governments ideally act only as facilitators, advisers and providers of guidelines, or at best as participants in co-management. However, in situations where the issue of unequal power relations has not been dealt with, as in the Model Forest Management Area in Southwest Bintulu in Sarawak (parts of which were located on land claimed under customary tenure), local management systems may not be accorded the respect required for participatory management. In this instance, according to Pedersen and his colleagues (Chapter 9), groups with diverging interests (loggers and local people) could 'co-exist positively' because of the initial economic gains that emerged from the presence of logging camps, including jobs, new fish ponds, and income-generation schemes such as pepper production. In the long term, there are predictable downsides and the ever-present potential for conflict over land between communities whose priority is to have sufficient land for subsistence agriculture and logging concessionaires who are interested in timber for profitable logging.

Clearly, participation based on accenting ethnic difference has had mixed results. However, analysts have warned against the potential risks of placing too much emphasis on difference.

Indonesia has a history of popular struggles that were phrased … not as claims of distinctive, culture-bound communities (*masyarakat adat*), but as struggles of 'the people' (*rakyat*). Is the shift of focus from people to culture, which coincides with a shift of the site of struggle from agricultural land to forests and nature, the best approach to justice? (Li 2003: 383).

This Volume

All chapters in this volume deal with the state, market and communities, but with differing levels of emphasis. Part II looks at the institutional framework that has effected dramatic changes to local histories, livelihoods and identities. The emphasis in this section is on the state and its institutions, including the law, as active agents of change. Although the effects of government policies on local communities are discussed, analysis of community strategies is not central, as it is in the second part of the book. Part III presents five case studies, three from Kalimantan (Deddy, Eghenter and Wadley), one from Sarawak (Pedersen et al.) and one from Sabah (Vaz). The case studies presented by Eghenter, Vaz and Deddy are written from the perspective of field workers who were directly involved in experiments with alternative development strategies. In summary, these attempts included building community capacity to a level where local groups could gain recognition as legitimate partners in the management of a conservation area (Eghenter), gaining recognition for local institutions and management systems through land titling (Vaz), and the practice of community mapping (Deddy). On the other hand, in their capacity as observers, Wadley, Pedersen, Mertz and Hummelmose describe the domination of logging interests in local political economies and the different ways in which local communities have dealt with this domination.

References

Agarwal, A. and C.G. Clark, 1999. 'Enchantment and Disenchantment: The Role of Community in Natural Resource Conservation.' *World Development* 27(4): 629–649.

Benjamin, G., 2002. 'On Being Tribal in the Malay World.' In G. Benjamin and C. Chou (eds), *Tribal Communities in the Malay World: Historical, Social and Cultural Perspectives*. Leiden: International Institute for Asian Studies and Singapore: Institute of Southeast Asian Studies.

———, 2005. 'Indigeny and Exogeny: The Fundamental Social Dimension?' Paper presented to the Ethnography and Development Research Unit, University of Malaysia Sabah, Kota Kinabalu, 12 September.

Berkes, F. and C. Folke (eds), 1998. *Linking Social and Ecological Systems.* Cambridge: Cambridge University Press.

Brohman, J., 1996. *Popular Development, Rethinking the Theory and Practice of Development*. Oxford: Blackwell.

Brookfield, H. and Y. Byron, 1990. 'Deforestation and Timber Extraction in Borneo and the Malay Peninsula: The Record Since 1965.' *Global Environmental Change, Human and Policy Dimensions* 1: 42–56.

Brookfield, H., L. Potter and Y. Byron, 1995. *In Place of the Forest: Environmental and Socio-economic Transformation in Borneo and the Eastern Malay Peninsula*. Tokyo: United Nations University.

Brosius, P., 1997. 'Prior Transcripts, Divergent Paths, Resistance and Acquiescence to Logging in Sarawak, East Malaysia.' *Comparative Studies in Society and History* 39(3): 468–510.

————, 1999. 'Anthropological Engagements with Environmentalism.' *Current Anthropology* 40(3): 277–309.

Casson, A., 1999. *The Hesitant Boom: Indonesia's Oil Palm Sub-Sector in an Era of Economic Crisis and Political Change*. Bogor: Center for International Forestry Research.

Chala, T., 2000. The 'Green Gold' of Sabah: Timber Politics and Resource Sustainability. Melbourne: University of Melbourne (Ph.D. thesis).

Curran, L., 1999. 'Ecosystem Management and Forest Policy in Indonesian Borneo.' Paper presented at the symposium on 'Ecosystem Management for a World We Can Live In', School of Natural Resource and the Environment, University of Michigan, Ann Arbor, 25 September.

Direktorat Jenderal Perkebunan (DJP), 1990–2004. 'Statistik Perkebunan Indonesia: Kelapa Sawit [Indonesian Estate Crop Statistics: Oil Palm].' Jakarta: DJP.

Doolittle, A., 2001. 'From Village Land to "Native Reserve": Changes in Property Rights in Sabah, Malaysia 1950–1996.' *Human Ecology* 29(1): 69–98.

————, 2004. 'Powerful Persuasions: The Language of Property and Politics in Sabah 1881–1996.' *Modern Asian Studies* 38(4): 821–858.

Eldridge, P., 1996. 'Human Rights and Democracy in Indonesia and Malaysia: Emerging Contexts and Discourse.' *Contemporary Southeast Asia* 18(3): 298–319.

Ellen, R., P. Parkes and A. Bicker (eds), 2000. *Indigenous Environmental Knowledge and its Transformations*. Amsterdam: Harwood.

Escobar, A., 1995. *Encountering Development: The Making and Unmaking of the Third World*. New Jersey: Princeton University Press.

Ferguson, J., 1990. *The Anti-politics Machine: 'Development', Depoliticization, and Bureaucratic Power in Lesotho.* Cambridge: Cambridge University Press.

Filer, C. (ed.), 1997. *The Political Economy of Forest Management in Papua New Guinea.* Port Moresby: National Research Institute/ London: International Institute for Environment and Development.

Fisher, R.J., 2000. 'Poverty Alleviation and Forests: Experiences from Asia.' Paper prepared for the pre-congress workshop on 'Forest Ecospaces, Biodiversity and Environmental Security', World Conservation Congress, Amman (Jordan), 5 October.

Friedmann, J., 1992. *Empowerment: The Politics of Alternative Development.* Oxford: Blackwell.

Howell, J. and J. Pearce, 2001. *Civil Society and Development: A Critical Exploration.* Boulder (CO) and London: Lynne Rienner.

ITTO (International Tropical Timber Organisation), 1990. 'The Promotion of Sustainable Forest Management: A Case Study in Sarawak, Malaysia.' Report submitted to the Eighth Session of the International Timber Council, Bali, 16–23 May.

Kalimantan Barat, 2003. 'Kalimantan Barat Dalam Angka [West Kalimantan in Figures].' Pontianak: Kantor Statistik [Provincial Statistical Office].

Kalimantan Selatan, 2003. 'Kalimantan Selatan Dalam Angka [South Kalimantan in Figures].' Banjarmasin: Kantor Statistik [Provincial Statistical Office].

Kalimantan Tengah, 2003. 'Kalimantan Tengah Dalam Angka [Central Kalimantan in Figures].' Palangkaraya: Kantor Statistik [Provincial Statistical Office].

Kalimantan Timur, 2003. 'Kalimantan Timur Dalam Angka [East Kalimantan in Figures].' Samarinda: Kantor Statistik [Provincial Statistical Office].

Khan, A., 2001. 'Preliminary Review of Illegal Logging in Kalimantan.' Paper presented at the conference on 'Resource Tenure, Forest Management and Conflict Resolution: Perspectives from Borneo and New Guinea', Australian National University, Canberra, 9–11 April.

Khor, K.P., 1993. 'Reforming North Economy, South Development, and World Economic Order.' In J. Brecher, J. Brown Childs and J. Cutler (eds), *Global Visions: Beyond the New World Order.* Boston: South End Press.

Kollert, W., L.K. Koon, M. Steel and S.M. Jensen, 2003. 'Biodiversity Conservation in Sabah's Commercial Forest Reserves — Status and Development Options.' Kota Kinabalu: Report to the Sabah Wildlife Department and DANIDA Capacity Building Project.

Li, T.M., 1999. *Transforming the Indonesian Uplands*. Amsterdam: Harwood/Singapore: Institute of Southeast Asian Studies.

————, 2001. 'Relational Histories and the Production of Difference on Sulawesi's Upland Frontier.' *Journal of Asian Studies* 60(1): 41–66.

————, 2003. '*Masyarakat Adat*, Difference, and the Limits of Recognition in Indonesia's Forest Zone.' In D.S. Moore, J. Kosek and A. Pandian (eds), *Race, Nature and the Politics of Difference*. Durham (NC) and London: Duke University Press.

MacLean, K., 2000. 'Constructing "Civil Society": Assessing Participatory Development in Contemporary Vietnam.' In G. Hainsworth (ed.), *Globalisation and the Asian Economic Crisis, Indigenous Responses, Coping Strategies and Governance Reform in Southeast Asia*. Vancouver: University of British Columbia.

Majid Cooke, F., 1999. *The Challenge of Sustainable Forests: Forest Resource Policy in Malaysia 1970 to 1995*. Sydney: Allen and Unwin/Hawaii: University of Hawaii Press.

————, 2002. 'Oil Palm and Vulnerable Places in Sarawak: Globalisation and a New Era?' *Development and Change* 33(2): 189–201.

————, 2003a. 'NGOs in Sarawak.' In M.L. Weiss and S. Hassan (eds), *Social Movements in Malaysia: From Moral Communities to NGOs*. London and New York: Routledge Curzon.

————, 2003b. 'Maps and Counter-maps: Globalised Imaginings and Local Realities of Sarawak's Plantation Agriculture.' *Journal of Southeast Asian Studies* 34(2): 265–284.

Malaysian Palm Oil Board, 2004. 'Malaysian Oil Palm Statistics 2004.' Viewed 14 September 2005 at http://www.mpob.gov.my/

Mannan, S., 1998. 'Sustainable Forest Management in Sabah.' In F. Kugan, E. Juin and P. Malim (eds), *Proceedings of the Seminar on Sustainable Forest Management*. Sandakan: Sabah Forest Department.

Nelson, P., 1995. *The World Bank and Non-Governmental Organizations: The Limits of Apolitical Development*. London: MacMillan/New York: St Martin's Press.

Nicholson, D.I., 1979. 'The Effects of Logging and Treatments on the Mixed Dipterocarp Forests of South East Asia.' Rome: Food and Agriculture Organisation of the United Nations.

Nussbaum, R., 1995. 'The Effects of Selective Logging of Tropical Rainforest on Soil Properties, and Implications for Forest Recovery in Sabah.' Exeter: University of Exeter.

Peluso, N., 1992. 'The Ironwood Problem, (Mis)management and Development of an Extractive Reserve Forest Product.' *Conservation Biology* 6(2): 210–219.

————— and P. Vandergeest, 2001. 'Genealogies of the Political Forest and Customary Rights in Indonesia, Malaysia, and Thailand.' *Journal of Asian Studies* 60(3): 761–812.

Potter, L., 1991. 'Environmental and Social Aspects of Timber Exploitation in Kalimantan, 1967-89.' In J. Hardjono (ed.), *Indonesia: Resources, Ecology and Environment*. Kuala Lumpur: Oxford University Press.

—————, 2005a. 'Commodifying, Consuming and Converting Kalimantan's Forests, 1950–2002.' In P. Boomgaard, D. Henley and M. Osseweijer (eds), *Muddied Waters: Historical and Contemporary Perspectives on Management of Forests and Fisheries in Island Southeast Asia*. Leiden: KITLV.

—————, 2005b. 'The Oil Palm Question in Borneo.' Paper presented at a conference on 'The Heart of Borneo', Leiden, 25–28 April.

Rahnema, M., 1992. 'Participation.' In W. Sachs (ed.), *The Development Dictionary*. London: Zed Books.

Ross, M.S., 1984. Forestry in Land Use Policy for Indonesia. Oxford: University of Oxford (Ph.D. thesis).

—————, 2001. *Timber Booms and Institutional Breakdown in Southeast Asia*. Cambridge: Cambridge University Press.

Sarin, M., 1999. 'Policy Goals and JFM Practice: An Analysis of the Institutional Arrangements and Outcomes.' New Delhi: Worldwide Fund for Nature (Policy and Joint Forest Management Series 3).

Scott, J.C., 1998. *Seeing Like a State: How Certain Schemes to Improve the Human Condition Have Failed*. New Haven (CT): Yale University Press.

Shiva, V., 1993. 'The Greening of the Global Reach.' In J. Brecher, J. Brown Childs and J. Cutler (eds), *Global Visions: Beyond the New World Order*. Boston: South End Press.

Smith, J., K. Subardi Obidzinski and I. Suramanggala, 2003. 'Illegal Logging, Collusive Corruption and Fragmented Governments in Kalimantan Indonesia.' *International Forestry Review* 5(3): 293–302.

Tinal, U. and J.L. Palenewen, 1978. 'Mechanical Logging Damage after Selective Cutting in the Lowland Dipterocarp Forest at Baloro, East Kalimantan.' In S. Rahardjo et al. (eds), *Proceedings of the Symposium on the Long-Term Effects of Logging in Southeast Asia*. Bogor: BIOTROP (Special Publication 3).

Weiss, M.L., 2003. 'Malaysian NGOs: History, Legal Framework and Characteristics.' In M.L. Weiss and S. Hassan (eds), *Social Movements in Malaysia. From Moral Communities to NGOs*. London and New York: Routledge Curzon.

Part II. Framework and Institutions

Chapter Two

Expanding State Spaces Using 'Idle' Native Customary Land in Sarawak[1]

Fadzilah Majid Cooke

It is noted that claims over *pulau* ... could not be sustained for reasons that the characteristic of a *pulau* ... is that it is a small pocket of original jungle deliberately preserved by the natives, i.e. ... it remains a virgin jungle. To acknowledge native customary rights ... would not be consistent with the cardinal principle that for the creation of NCR ... a native must clear the land for farming and remain in occupation thereof (Fong 2000: 18).

Introduction

Current interest in the decentralisation of state and administrative power has provided lessons about state strengths or weaknesses and why the reform process in many countries has met with difficulties. Examining factors contributing to those difficulties by studying state management of natural resources could provide a beginning for understanding the challenges faced by reformists.

Following Dove (1986, 1999), the state is seen here as having its own developmental and environmental agenda, but is not monolithic. A most ambitious social engineering program has been attempted in Sarawak, East Malaysia, since the mid-1990s, with the large-scale redesign of rural life through the introduction of plantation agriculture. This chapter argues that oil palm development from the mid-1990s and continuing into the 21st century is different from that of earlier decades in the systematic targeting of 'Native Customary Land', or land claimed under native customary rights.[2] The systematisation is reflected in discourses and practices concerning the management of land and

---

[1] This chapter has been reshaped from an earlier paper presented at the Resource Management in Asia-Pacific Program conference on 'Resource Tenure, Forest Management and Conflict Resolution: Perspectives from Borneo and New Guinea', held at the Australian National University, Canberra, 9–11 April 2001. Parts of the paper presented at that conference, combined with others presented elsewhere, were published in the journal *Development and Change* (Majid Cooke 2002). Changes occurring on the ground in Sarawak since 2002 have influenced the tone and content of the present chapter. The last section is based on previously unanalysed fieldwork material in the middle Baram and Ulu Teru. Fieldwork was conducted in April–May 2000 and August–September 2001 and formed part of an ongoing interest in Sarawak beginning 12 years ago.

[2] In some discourses, official or otherwise, Native Customary Land is also referred to as 'Native Customary Rights Land'.

the Dayak peoples of Sarawak.[3] Differing from views that regard this kind of development as merely bringing Dayak peoples into the 'mainstream' of economic life, this chapter suggests that oil palm development under Konsep Baru (New Concept) is concerned with expanding state spaces.[4] 'Contemporary development schemes, whether in Southeast Asia or elsewhere, require the creation of state spaces where the government can reconfigure the society and economy of those who are to be "developed"'(Scott 1998: 185).

The expansion of state spaces involves a range of strategies. One frequent feature of such strategies is that they reflect a normative 'civilising process' (Scott 1998: 184). First, the process promotes the depersonalisation of social life and its separation from economic life. Once depersonalised, social life can then be conceived of solely in *economic* terms.[5] In Sarawak, Dayak groups are frequently warned about their 'backwardness' and are regularly informed of their rights as economic citizens; in other words, their 'right to develop'. Expressions of other rights (such as cultural or political rights) by citizens are regarded as venturing outside the realm of citizenship, or a result of prompting by 'unscrupulous elements' or 'trouble makers'.[6]

A second strategy of state expansion involves territorialisation (Vandergeest 1996). In this process, state power is expanded into local geographies and economies through administration, with legal codes and classification systems put in place to enable the state to take over local property rights. Vandergeest and Peluso (1995: 385) have argued that the state exercises power in actions that 'include or exclude people within particular geographic boundaries', which 'control what people do and their access to natural resources within those boundaries'. The state's territorial organisation of people and economic activities makes use of abstract space (guided by maps and land use planning) which often

---

[3] Since this is not a paper focusing on ethnicity, 'Dayak' here refers to all the ethnic groups in Sarawak who are largely non-Muslim, especially Iban, Bidayuh and Orang Ulu. Among the more prominent of the Orang Ulu group are the Kayan, Kenyak, Kelabit, Lun Bawang and Penan.

[4] State spaces in Scott's (1998: 186–9) conceptualisation are best understood in relation to non-state spaces. State spaces are those localities that are economically and politically visible to the state and largely controlled by it. In state spaces, economic surplus is usually generated through state-managed development programs, often involving discipline or appropriation of resources. Being regarded as 'civilised', state spaces are integral for the extraction of economic surplus and labour, and populations are disciplined through codification of their religion, settlements and households. By contrast, non-state spaces are those in which are found populations that cannot be relied upon to produce all those surpluses that the state requires. Such spaces and their inhabitants are regarded as 'exemplars of rudeness, disorder and barbarity' and, more worryingly for the state, have often 'served as refuges for fleeing peasants, rebels, bandits, and … pretenders' (Scott 1998: 187).

[5] For a review of the literature on economic nationalism and the creation of Malaysian citizenship in economic terms, see Williamson (2002).

[6] Daily newspapers are full of these expressions. Some examples include: 'Taib: Interaction vital at grassroots level to create greater understanding' (*Sarawak Tribune*, 11 May 2000); 'Taib takes tough stance against trouble-makers' (*Sarawak Tribune*, 11 May 2000); 'Tajem: If a Dayak cannot speak for Dayaks, who can?' (*Daily Express*, 21 June 2004); 'Explain NCR land development policy, YBs told' (*Daily Express*, 14 July 2004). (YB is Malay for *Yang Berhormat*, and is a local form of address for elected State Assembly members.)

does not correspond with people's lived space (ibid.: 385–6). This takeover process is best exemplified when viewed historically (Peluso and Vandergeest 2001).

This chapter argues that the current attempt in Sarawak at providing economic value to Native Customary Land through a form of land certification is based on what I have termed a fundamental error, arising from a misinterpretation of unoccupied land as 'idle' or 'waste' land, originating during the Brooke period in the 19th century and continuing into colonial times between 1945 and 1963. This error resulted in serious repercussions for local access and management regimes, and has still not been questioned today. In contemporary times, and in association with Konsep Baru, the introduction of the *Land Code Amendment* of 2000 further perpetuates this error. It is important to examine the basis and usefulness of the error in order to understand why it has been perpetuated.

However, the state is not monolithic, and in the context of natural resources, although states may have coercive power, complete control cannot be assumed (Rangan 1997). As well, tension emanates from the imperatives for control and the need for state legitimacy. This tension led to the 'discovery' of customary rights in colonial times and of 'development' today. Because development in Sarawak, as in Peninsular Malaysia, is directed not only towards physical or structural change, but towards cultural transformation as well, contestation in the cultural realm is possible. Living at the frontier as they do, some of the people who are regarded as requiring 'civilising' can have a different view of development. For many the creation of state spaces is traumatic; for others the process is tolerable. But for all, some amount of 'persuasion' is required. From its inception in the mid-1990s until 2005, only 17 per cent of the Native Customary Land targeted for development under Konsep Baru has been successfully converted, prompting the Sarawak Assistant Minister for Land Development to exclaim that '[we] have only five more years to go to achieve our target' (*Daily Express*, 7 September 2005).

Focusing on strategies employed by the state in promoting oil palm development, this chapter discusses the methods involved in state 'persuasion' processes associated with the introduction and implementation of Konsep Baru, and also the limits of these strategies. Attempts at expanding the public space from below have been treated elsewhere (see Majid Cooke 2003a, 2003b).

High Modernism in Sarawak

Malaysia is a developmentalist state (Embong 2000), and Sarawak lives up to Scott's description of high modernism (Scott 1998). Developmentalist states assume a direct role in promoting and guiding economic expansion and growth. High modernist ideology generally favours rational engineering of entire social orders in creating 'realisable utopias', pervasive planning and rationalised production (Scott 1998: 97–8). In Sarawak, strategies used by political parties in power, bureaucracies, and non-state economic actors make two assumptions: that rural Dayak groups are vulnerable to being left behind in the face of globalisation, and that they need to be brought into the 'mainstream' of development (Majid Cooke 2002).

As a land development program, Konsep Baru is viewed by officials as promoting Dayak into the 'mainstream' of economic development. It promotes the conversion of Native Customary Land into oil palm plantations. Although plantations are not new to Sarawak, having been introduced by the government since the 1960s (Ngidang 2002), Konsep Baru is novel in its systematic targeting of land claimed under customary rights. That the emphasis is on productivity is clear. Lands regarded as 'idle' are to be converted into tangible assets, in the form of shares, so that native peoples can become shareholders in oil palm companies working on their land on a joint venture basis. The ratio for the joint venture is 60 per cent for the company and 30 per cent for local communities, with the government acting as a trustee and enjoying a 10 per cent share. Companies are given provisional leases of 60 years (considered good for two production cycles and a necessary incentive for return on investment), the renewal of which will be dependent on the outcome of negotiations among stakeholders. Local communities as landowners are to receive 'certificates of title' upon registration of their land in a land bank. Such registration enables the conversion of rural landscapes into oil palm plantation blocks (see Figure 2.1).

To help hasten the registration of Native Customary Land for use in the joint ventures, the *Land Code Amendment* was introduced in the year 2000. The emphasis on productivity is different from earlier phases of plantation development which had a mixture of aims, one of which was Dayak social and economic development on what was considered in planning circles as 'state land' (Ngidang 2002). Community responses to Konsep Baru have been varied, ranging from outright acceptance to resistance (Majid Cooke 2002). In between are those who engage in strategic agriculture, converting their land to oil palm themselves, as a smallholder crop, ahead of company bulldozers. Overall, regardless of the responses, there has been widespread anxiety over many issues, including: the exact role of government as trustee, if and when the oil palm market crashes; the tying up of land for 60 years; and confusion over what 'certification' implies, that is, whether it suggests individual or group entitlements and what it means in terms of access in the long term (see Majid Cooke 2002, 2003a).

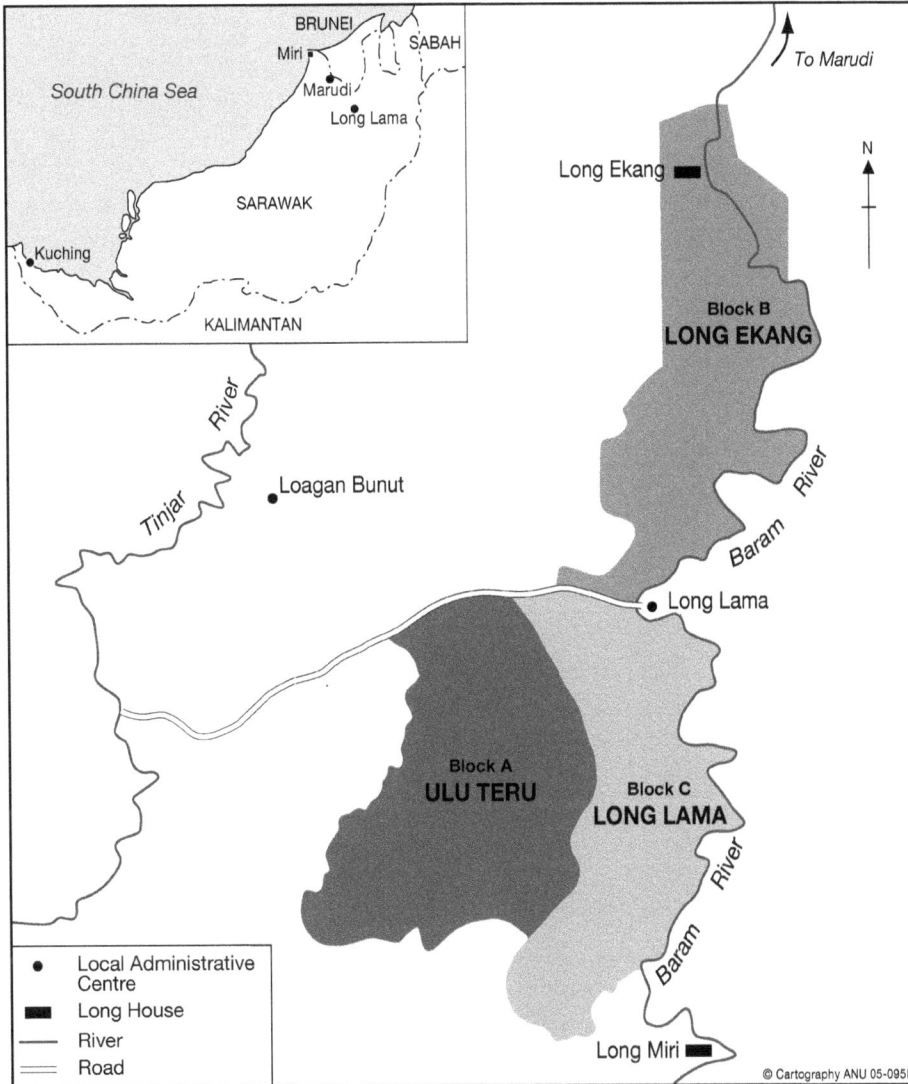

Figure 2.1. Plantation blocks of the middle and lower Baram, Sarawak

These concerns are the surface manifestations of a much deeper uncertainty over tenure. Although customary rights are legally acknowledged in the Sarawak *Land Code*, they are often superseded or ignored in practice, depending on which state institution interprets them. While there are moves towards making customary rights universal through the court system, in most interpretations state institutions regard customary rights as contingent, easily overridden by state development needs (see Majid Cooke 2003a). A major question was how land 'certificates' could provide customary owners with better security against future capture by government development projects or against potential manipulation by large corporations (Majid Cooke 2002).

A second group of concerns was linked to the ambitious scope of the project. Although targets were not specific, officials made clear that, in order to meet the target of creating one million hectares of oil palm by 2010, at least 400 000 hectares were to come from Native Customary Land (interviews at Kuching, September 2001; *Daily Express*, 14 July 2004). Since many issues remained unresolved, the ambitious scale of the program created anxiety. Unresolved doubts about the status of the land have not died down, and they became an election issue in the Ba Kelalan by-election in October 2004, providing the opposition candidate with a substantial margin of votes (*Daily Express*, 20 September and 3 October 2004).

As recently as 2004 and 2005, many issues remained unresolved, while others had surfaced. These were raised during a number of hearings organised by the Malaysian Human Rights Commission. At the hearings, questions were raised about administrative and planning procedures, and interpretations made about government 'sincerity' with regard to developing Native Customary Land (*Daily Express*, 31 December 2004, 1 February 2005). First, in relation to administrative and planning procedures, questions were raised by rural Dayak as to whether frequent encroachments onto their land were a result of a planning process where maps were drawn awarding provisional leases or logging concessions without appropriate ground checks being made. Second, with regard to the many concerns among rural Dayak about land ownership, a key issue was whether land 'certificates' would be given to individuals or to 'landowning groups'. Many answers were given in the meantime. Finally, when some Dayak landowners drew their own conclusion, they were said to be sceptical about the government's real intention in land development because, since the adoption of the *Land Code Amendment* pushing for land registration under Konsep Baru, they had observed a lack of government action to make Native Customary Land more transparent. Since 2000, Dayak claimed that progress on perimeter surveys of customary land had been slow, as was the process of issuing titles to owners (*Daily Express*, 1 March 2005). As a result, they claimed to have experienced continued encroachment onto their land by logging and oil palm companies. Among those who have submitted their land for development under Konsep Baru, a new concern is the temptation of selling their developed land for quick gain. Using the sale or attempted sale of titled customary land by some Dayak individuals at Kanowit (one of the first areas where customary landowners enlisted in the Konsep Baru program), the government expressed concern about further titling (*Daily Express*, 17 May 2004). When such sales take place, Dayak are considered innocent objects, 'robbed' by reckless speculators, not a people who actively made viable decisions considering the new and complex situations they find themselves in, even though some of those involved were said to be a syndicate of 'prominent community leaders and businessmen' (*Daily Express*, 18 and 27 May 2005). With many issues remaining unresolved. Konsep Baru is seen to be moving slowly.[7]

Officials and planners attribute Dayak selectiveness in embracing development projects to traditionalism or anti-developmentalism. Under Konsep Baru, native peoples are urged to become 'modern'. Modernity is contrasted with conservatism and this duality is seen as seeping through every level of social life. But it is the mental attitudes that officials want to change. 'A radical mental revolution is required to effect paradigm shift in the attitudes and perceptions of landowners towards developing [Native Customary Land]' (Sarawak Ministry of Land Development 1997: 17). Accordingly, to become 'modern', a person must see land as a commodity or an economic asset, to be traded and not treated as an heirloom nor seen as the only form of wealth. Dayaks are told that individuals must be prepared to take risks, share the cost of development (against being overly dependent on government assistance), and choose better alternatives for land use instead of keeping land in its 'original' state (ibid.: 17–18). In Sarawak persuasion is the key to the exercise of state power.

The process of persuasion as engaged in by state officials forms the crux of the next section of this chapter, which examines the discursive and practical element to Konsep Baru. Persuasion involves a two-step process. The first step entails convincing the more sceptical rural Dayak groups that they are in danger of being left behind, and that Konsep Baru is a vehicle for 'catching up'. A second step calls for persuading rural groups to accept a trade-off: economic development and welfare against some aspects of citizenship rights, especially freedom of expression and association. Given spatial limitations, the second step will be dealt with only briefly (based on fieldwork in Ulu Teru in 2000 and 2001), and ought to be the subject of another paper.

Persuasion: Creating Vulnerable Identities and Places

For at least two years prior to the adoption of Konsep Baru in 1994, Dayak 'vulnerability' was a recurring theme adopted in different guises in the print media. Rural communities had real concerns over major shifts in their status from landowners to workers or minor shareholders in plantation companies.

---

[7] Dayak scepticism required assurance from political leaders: 'Large scale development is not a ploy to take away land from the owners but a genuine long-term plan to bring them into the development mainstream' (Deputy Chief Minister of Sarawak, quoted in the *Daily Express*, 6 July 2005); 'We want to use especially Native Customary Rights (NCR) lands for the purpose [of large scale plantation projects] because we want to help the native landowners' (the Chief Minister of Sarawak, quoted in the *Daily Express*, 30 June 2005). The semi-government agencies responsible for the implementation of Konsep Baru reported that in order to achieve the set targets they had to work hard, but their objectives are far from being met. As of September 2005, only 70 000 of the targeted 400 000 hectares (17 per cent) of Native Customary Land have been converted. Since the deadline for achieving this goal was the year 2010, the Sarawak Assistant Minister for Land Development expressed his concern, '...and [we have] only five more years to achieve our target' (*Daily Express*, 7 September 2005).

Fundamental concerns about access to the means of livelihood and tenure were not addressed or, if they were, only handled in a negative manner. These concerns were considered unfounded or old-fashioned by the proponents of Konsep Baru, who painted economic globalisation as inevitable and the perceived Dayak backwardness (*katak di-bawah tempurong* — 'frog under the coconut shell') as making them more 'vulnerable'. If rural Dayak did not become active participants in globalisation, it was claimed, they might be left behind: 'We have to transform the rights to our land into much more tangible assets that can increase in value, that can be transferable, and that can fit into our system of trade and business. Otherwise we will be left out.'[8] Public debates were reduced to issues regarding technicalities, such as appropriate measures to adopt for land registration, so as to avoid confusion or delay, and for just compensation when land is withdrawn from the people for development.[9]

Official claims of Dayak vulnerability are based on perceptions about their land tenure system, of leaving rich land 'idle', as the following quotations show. Chief Minister Taib was quoted in the *Borneo Post* of 19 July 1999 as saying: 'The NCR [Native Customary Rights] land will be of not much use, unless something productive is done to exploit the natural wealth to help the people.' He was supported by Polit Hamzah, General Manager of the Land Consolidation and Development Authority:

> A large number of people have lands but they do not do anything about it. The Chief Minister wants to transform these lands into tangible assets. If these abandoned lands are transformed into plantations, landowners will receive shares making their rights over the lands become tangible assets (quoted in *New Reality*, May–June 2000: 18–19).

Similarly, an industry representative spoke of native peoples simply 'wandering around' in the jungle because there was abundant land.[10]

Dayak can circumvent this vulnerability to poverty by making their land more marketable — so the argument goes. One sure way of making 'idle' land (*tanah terbiar*) more productive is through plantation agriculture. Plantations are equated with progress, while progress is seen as inevitable and not to be prevented: 'Land owners who take harsh action to prevent the government's

---

[8] Chief Minister Taib, quoted in the *Borneo Post*, 8 March 2000. See also: 'Taib raps instigators of false land claims' (*Borneo Post*, 19 May 1999); 'Amendment to Land Code: best Gawai gift to natives' (*Borneo Post*, 23 May 2000).

[9] See, for instance: 'Pendaftaran NCR hapus keraguan raayat' (*Utusan Borneo*, 11 May 2000); 'Landowners should be given share options, says Lee' (*Borneo Post*, 11 May 2000); 'MP approve genuine NCR land directly' (*Borneo Post*, 11 May 2000); 'Amendment to Land Code, a jewel in BN's Crown' (*Borneo Post*, 11 May 2000); 'Govt. sincere in recognising rights of native over NCR land' (*Sarawak Tribune*, 11 May 2000).

[10] Chairman, Perlis Plantations, at the seminar on 'Undang-Undang dan Pembangunan Tanah Zon Utara (Law and Land Development in the Northern Zone)', Sibu, 1999.

move to develop their land through the private firms will only mar their own progress' (Chief Minister Taib, *New Reality*, May–June 2000: 16).

The theme of rich but 'idle' land is picked up by many in government,[11] and is shared by some elements within the public service, although among the latter there is a range of views. From one point of view, the 'complex land tenure problem' see is expected to melt away as farmers become better educated and are drawn away from farming activities (Chua 1992: 104). On the other hand, the position represented by Dandot (1992) suggests that neither 'idle' land nor the land tenure system is to blame, but rather: 1) the failure to provide clear explanations about the benefits or options of land development that are available to farmers; and 2) the inadequate surveying, demarcation and recognition of customary land prior to the introduction of land development schemes. However, both positions regard the native land tenure system as a 'problem' to be solved, not a solution for arriving at alternative types of land development.

Recently, the ideology of rural backwardness and vulnerability has not diminished nor taken on new form. In 2004, close to 10 years after the inception of Konsep Baru, the critique against rural cultural life as embodied in notions about 'idle' or 'waste' land continues.[12]

Of 'Idle' or 'Waste' Land

All over southeast Asia, viewing unoccupied customary land as 'idle' or 'waste' land appears in different guises, and is based on a fundamental error which has several features. First, colonial legal codes transformed forests into two categories: 'natural forests' (a political category) and 'agricultural land'. Local management systems do not differentiate between the two categories. Once local spaces have been transformed through categorisation, they are then policed using techniques of power and discipline that include territorial zoning and mapping, the constitution of institutions of enforcement, and the creation of exemptions, among which are customary rights. '[T]he creation of Customary Rights and reference to this political process as "discovery" or "recognition" allowed state actors ... to appear generous in conceding access' (Peluso and Vandergeest 2001: 765).

During the Brooke period, ideologies of legal pluralism as well as financial constraints limited the space taken over as 'state land', so that the tension caused by the recognition of 'customary rights' and expanding state spaces was avoided.

---

[11] Some examples from the popular media include: 'Amendment to Land Code: best Gawai gift to natives' (*Borneo Post*, 23 May 2000); 'Third plantation on Native Land' (*Borneo Post*, 2 March 1997); 'Kerajaan di-gesa wajibkan syarikat balak bantu pendudk terjejas' (*Borneo Post*, 13 May 2000). See next section of this chapter for the historical origins of the concepts of 'waste' and 'idle' land in Sarawak.
[12] 'Explain NCR land development policy, YBs told' (*Daily Express*, 14 July 2004); 'Owners of NCR land can make RM 5000: Chin' (*Daily Express*, 23 August 2004).

In the colonial period, the tension re-emerged and is seen in the ambivalence about customary land in the 1958 *Land Code*.

The provision for recognition of customary rights in the 1958 *Land Code* allowed occupation prior to 1958 by methods prescribed in Section 5(2) of the code.[13] The basis for the 1958 *Land Code* dates further back, to the Brooke period. Recognition of local people's pre-existing rights in land and trees can be seen in the *Code of Laws* introduced in 1842, which was meant to forbid interference with native customary law and protect native peoples from 'immigrant races', especially Chinese (Peluso and Vandergeest 2001: 779). However, the 1863 *Land Regulations* gave the Brooke regime rights over all 'unoccupied and waste lands' which it could then lease out to individuals and companies (Richards 1961; Porter 1967).[14] 'Unoccupied and waste lands' were defined as those lands outside of territories where natives could hold 'customary land rights' (Peluso and Vandergeest 2001: 779, citing the 1863 *Land Order*).

This is the start of the problem for native communities because 'unoccupied and waste lands' covered all land regarded as lying outside those classified as customary land. Areas that may appear to be lying 'outside' customary land from the official perspective may in fact lie 'within' it from a local perspective. Later orders showed that there was no understanding among Brooke administrators about land left to fallow or rotation for shifting cultivation, or deliberately left uncultivated for ecological reasons (watershed protection), or subsistence use (for wild meat, building material, rattan supply for mat and basket making, and so on).[15] For example, in 1875 an administrative order was issued for the purpose of imposing a fine against the act of clearing land and then 'abandoning' it. 'This (administrative) Order ... suggests a curious misunderstanding on the part of Government, not simply of the practices permitted under native customary law but also of the biological demands the practices made on the land' (Porter 1967: 37).[16] This initial 'misunderstanding' regarding 'unoccupied and waste lands' continues to cause problems to native communities today, because the error remained uncorrected and unquestioned

---

[13] There were six methods: a) the felling of virgin jungle and the occupation of the land thereby created; b) the planting of land with fruit trees; c) the occupation or cultivation of land; d) the use of land for a burial ground or shrines; e) the use of land of any class for rights of way; or f) any other lawful method.

[14] On the encounter between native agriculture and colonial administration over the issue of 'unoccupied and waste land' in another context, see also Leach and Mearns (1996).

[15] There is a whole range of practices with similarities and differences in land tenure systems in Sarawak, as there are in the rest of the island Borneo (see Appell 1997). On joint community reserves for the Iban in southern Sarawak see Cramb (1989), Rousseau (1987) and Sather (1990). In Saribas, where Cramb worked, the reserved forest area included land strips bordering the Layar River which had fruit trees and illipenut trees planted many years previously, as well as the sites of many old longhouses (*tembawai*) (Cramb 1989: 282–3).

[16] The imposition of the 1863 *Land Regulations* 'meant that indigenous groups could no longer automatically acquire additional land rights by clearing forest outside their existing territory' (Cleary and Eaton 1996: 55).

in the post-colonial era. The problem can be glimpsed in the quotation from the current Sarawak State Attorney General cited at the beginning of this chapter.

The idea that land was productive only when occupied or improved went hand in hand with legalising land ownership through title. These notions reflect elements of the Torrens system of property entitlement that influenced the Brooke land policies of the 1930s (Porter 1967: 51). During the Brooke era (and especially from 1875 onwards) a number of ideas were introduced: that natives were 'squatters' on government or 'state-owned' land; that rights were tied to specific lots in 'native land reserves'; and that surveillance was necessary to control cash crop production (ibid.: 35–51). Cash crop production was to be controlled through species regulation (gambier, rubber and pepper were encouraged) and through titles (registration, permits, leases or occupation tickets). Surveillance of species and territory was facilitated in the 1930s, when 'fairly accurate maps' were printed and published for the first time (ibid.: 49).

During the colonial period (1946–63), a new system of classification was introduced through the 1948 *Land (Classification) Ordinance* which divided land into five categories: mixed zone, native area, native customary, reserved, and interior area land. The design of the classification system was racialised in that it was intended: 1) to protect natives from encroachment by non-natives, by restricting the latter to land within the 'mixed zone'; and 2) to prevent natives from disposing of their land to non-natives (Porter 1967: 62–3). However, these intentions need to be framed against plans to open up Sarawak's forests to logging on a far larger scale than had been done during the Brooke era (Majid Cooke 1999: 46), which would have severe implications for customary land, given the lack of capacity of the colonial land surveying machinery to ensure sufficient safeguards. Equally, during the colonial period, the 1958 *Land Code* was found to be ineffective in terms of its original intention of 'protecting native interests', so that in 1962 a Land Committee proposed that tighter control be exercised through the Resident's office to prevent the 'disposal' of native land to non-natives.

Native Customary Land and 'Development'

State control over land did not change significantly in the post-colonial era, except that with intensified 'development' (state-sponsored oil palm in the 1960s, logging in the 1970s and 1980s, and then joint venture oil palm from the 1990s on), customary land came under severe pressure. When under pressure, and due to the ambivalence of the 1958 *Land Code*, 'state land' is often contested space. In brief, the *Land Code*, while acknowledging the rights of native communities to live on their land (access rights), was ambivalent about recognising their 'ownership' of the land. Today, this ambivalence allows for multiple interpretations of the status of Native Customary Land in relation to other lands managed by the state. For example, using one set of evidence

(Appendix to the *Native Customary Laws*), one interpretation of the law suggests that 'all untitled land whether jungle or cleared for padi farming (*temuda*) became "the property of the Crown"' (Fong 2000: 9). This is the view underpinning the quote by Fong at the beginning of this chapter. Without evidence including maps, records kept at district offices, or certificates issued by colonial administrators or Brooke officials, Native Customary Land was taken to be 'crown land' and Dayak landowners were interpreted as being mere 'licensees' (ibid.) as in the colonial period, except that they are now 'licensees' on 'state' rather than 'crown' land (ibid.: 19). However, most native communities were unaware of government edicts, since they did not have access to government gazettes, and continued to create new settlements and claim customary rights, both before and after 1958, in line with *adat* (customary) law. Regardless of the ambivalence in the *Land Code*, from the mainstream legal viewpoint such communities are 'illegal squatters' on state land (ibid.: 19), and the majority of those who occupied land after 1958 find their status especially uncertain.

A different view suggests that what observers mistake for 'virgin jungle' may in fact be *pulau*. From this perspective, *pulau* are forest reserves especially set aside by native communities for essential items such as timber for house construction and building boats, jungle vegetables, rattan and other produce. They may also provide the hunting ground for the community and be important water catchment areas (Bian 2000: 23). This view is in line with the contemporary discovery of 'anthropogenic forests' that were formerly classified as 'virgin' forests untouched by human managers (Leach and Mearns 1996).

Ambivalence and multiple interpretations are problematic to the hypermodernist agenda. The *Land Code Amendment*, introduced in conjunction with Konsep Baru, can be viewed as a way of cleaning up unfinished business left by the colonial legacy. Officially, the amendment was introduced for the dual purposes of recognising 'genuine' customary rights claims over land through land registration, and preventing 'all forms of unlawful occupation of State land on the pretext that such occupation is allegedly based on *adat*' (Fong 2000: 19). Registration of Native Customary Land is now seen in official circles as an attempt to provide statutory recognition to holders of customary rights, who were otherwise 'legally bare licensees in occupation of State land without title' (ibid.). For land to be registered, however, the usual requirements apply; namely, that natives would first have to be considered legal occupiers of their land prior to 1958, or would need proof that they had acquired customary rights over their land. The onus of proof is on the claimant. Proving that they have legitimately acquired their rights will keep most communities busy, regardless of whether they settled the land before or after 1958. For those who had settled after 1958, however, acquiring evidence is almost impossible because permits were very rarely issued to native landowners after that date (interviews at Miri, May 2000).

Having gone through the complicated process of claiming their rights, communities may then wish to have their land registered so that, under Konsep Baru, they could exchange it for shares in the joint venture oil palm companies working on their land. But registration does not make native peoples landowners in the eyes of the law. It is merely a 'registration of ... rights, not a registration of any estate or proprietary interests in land' (Fong 2000: 24). Small wonder, then, that as late as 2004, judging from the many appeals made by officials for natives to register their land, many rural groups seemed to be hesitant about doing so.[17] Some resist oil palm plantations through blockades of company vehicles and/or government-linked surveying teams, or through the courts (Majid Cooke 2003a). In court, Dayak groups have been more successful in obtaining compensation for damage done to their land than for recognition of their customary claims to land.[18] Their overall reluctance may be one reason for the perceived slow rate of acceptance of Konsep Baru in some rural areas of Sarawak.

The intention in passing the *Land Code Amendment* was to eliminate loopholes emanating from the 1958 *Land Code*. As noted, the *Land Code* recognised native customary rights, albeit in a limited way. The method described in Subsection 5(2)(f) as that of acquiring rights by 'any other lawful method' was particularly useful for negotiating access rights for those who had settled in their areas before 1958, but who possessed no legally acceptable evidence or proof.[19] However, it also proved useful for those who occupied land after 1958 and who, in the eyes of the law, were 'illegal squatters' on 'state land'.

Some in the legal profession saw Subsection 5(2)(f) as a way of bringing *adat* (custom) into the legal system. According to Baru Bian (2000), *adat* establishes a bundle of rights and practices that are otherwise not captured by the *Land Code*. *Adat* carries an inherent flexibility in terms of land access and use as demographic or economic pressures change (Rousseau 1987; Cramb 1989; Sather 1990). Subsection 5(2)(f) provided the potential for capturing this flexibility, and by deleting this subsection, the amendment of 2000 restricts the potential for claiming access to land through *adat*. The effects are summarised by Bulan

[17] Some examples include: 'Explain NCR land development policy, YBs told' (*Daily Express*, 14 July 2004); 'Owners of NCR land can make RM 5,000: Chin' (*Daily Express*, 23 August 2004).
[18] However, the 2005 Court of Appeal decision in favour of the Temuan (Orang Asli) group of Peninsular Malaysia in recognising their 'communal ownership' of land is regarded by many observers as having a significant implication for the status of customary land in Sarawak (as well as Sabah).
[19] Some of the rights acquired before 1958 were registered with the respective district offices, but others were not. Many communities did not register their land even if they had settled before 1958 because of lack of access to relevant information (especially government gazettes), or failure to move swiftly to register their land within the period stipulated by the *Land Code*. Land use classification and planning, on the other hand, is guided by aerial photography taken in or prior to 1958, which is now considered insufficient to distinguish vegetation details or settlement types, i.e. whether or not they were established prior to 1958. For some villages such photographs are simply not available (Fong 2000). After 1958, native occupation of land outside existing group territories was controlled, requiring settlers to first acquire a permit from the relevant district offices. In reality, such permits were rarely given.

(2000: 19) thus: 'The deletion of "any other lawful means" under Section 5(2)(f) appears to have taken away every remnant of practical right of the natives to prove entitlement to customary land.' This means that the rights of those who settled on their land after 1958 and who, under Subsection 5(2)(f), could have negotiated for recognition of their rights under *adat*,[20] remain unresolved. In some villages, even among those who have records to prove their occupation as being prior to 1958, the fear of Konsep Baru stems from the notion that it is a mechanism for 'taking away' Dayak land, with the benefits from plantations going to other people (Majid Cooke 2003a).

Trade-offs and the Limits of Persuasion

As mentioned before, development in Sarawak is considered to be a 'gift' from the government to the people. On the ground, the gift is translated in the provision of development goods, accessible to those who support the political parties in power. In rural areas such support can be well rewarded with good infrastructure (roads, energy and water supply) and services (especially education and health). People are made to understand that wanting development means supporting political parties in power, so that the continued provision of development gifts is conditional on sustained support for these parties. Such support is regarded as the people's recognition of the government's sincerity. By extension, doubts about particular development practices are regarded as doubting the sincerity of the government, and are by implication an ungrateful act. Furthermore, not wanting to vote for the ruling political parties is interpreted to mean that people are 'anti-development', which is considered a sin in Sarawak's hypermodernist tradition. The dominant image that emerges, and one that is officially sanctioned, is that there is only one type of development (the modernisation variety), and only one institution able to undertake it (the government — which in the Malaysian context is synonymous with the ruling parties).[21] This image is often supported in practice when planned projects are reportedly withheld when citizens vote for opposition parties, as happened in the 2004 by-election for the seat of Ba Kelalan in northern Sarawak.[22]

In top-down development such as Konsep Baru, the avenue for expressing dissent or merely asking for clarification about the program is extremely limited. As noted above, asking or raising questions at whatever level carries a certain

[20] Under the 1958 *Land Code*, the settlement of land beyond natives' existing boundaries is generally not allowed unless a permit is issued for this purpose by the local district offices, under Section 10(4) of the code (Fong 2000: 18). In reality, however, such permits are rarely given (interviews at Marudi and Miri, May 2000).
[21] There are many examples. A selection includes: 'People thirsty for development' (*Daily Express*, 20 September 2004); 'Mawan: people want development' (*Daily Express*, 20 September 2004).
[22] In this election, a group of people from the village of Long Semado in the Ba Kelalan constituency, decided to vote for an Independent candidate, an action which they may pay dearly for as they are reportedly threatened with the withdrawal of a Rural Growth Centre project already planned for the area (*Daily Express*, 31 October 2004).

amount of risk. As a result, confusion and lack of information abound among the implementers of Konsep Baru as well as the longhouse communities (Ngidang 2002).

Ngidang (2002: 168) claims that the politics of implementation of Konsep Baru is based on co-optation, which is built on a 'psychology of consensus'. For fear of being labelled deviant, 'anti-government' or 'anti-development', community leaders in Ulu Teru and Kanowit (where he conducted his fieldwork) agreed to participate in the joint venture program.[23]

Under Konsep Baru, though, there is some small room to ask questions. From the perspective of officials or implementers,[24] there are two kinds of queries, although they may be similar in substance. From the perspective of officials what makes the questions appear different is their source (who does the asking). The first type is raised by individuals, groups or institutions considered 'friendly' to the government and its programs. The second type of questioning may come from those who have a history of voting for political parties other than the dominant ones, those who have connections with non-government organisations (NGOs), or those who are merely seeking clarification or have reservations about public policies that affect their lives. These groups are regarded as 'unfriendly' (Ngidang 2002: 169) and are often ignored, while officials show a preference for communicating and interacting only with those considered 'friendly'. By contrast, among the 'friendlies', difficult questions can be comfortably glossed over, because even among officials some of the legal and economic implications of Konsep Baru are not fully understood (Ngidang 2002). One significant feature of Ngidang's findings is that the politicised nature of land development has created a pattern of communication wherein, at the official and community levels, key questions about Konsep Baru concerning security of tenure of Native Customary Land and the economic viability of joint ventures were not being

---

[23] In other parts of Sarawak, such as Bekenu near Miri, Iban participate in Konsep Baru, even when they know very little about it, because they 'hope' that government will look after their interests (Majid Cooke 2002). In the 1980s, large-scale oil palm agriculture was introduced to Bekenu by a state-sponsored development agency, the Sarawak Land Development Board. At the time, the people who protested against part of their customary land being mapped as 'state land' were arrested and later released. Many benefited from not opposing the project, and by the year 2000, were clearly being showered with development goods; namely, good roads, telephone connections, and some private and public sector employment in the nearby town of Miri for the younger generation. This time under Konsep Baru, Iban of Bekenu continued to put their trust in government, since they thought that the government, through its 10 per cent share in the joint venture program, would be involved in looking after its own interest, in addition to exercising its role as trustee.

[24] Officials appointed to implement the Konsep Baru are mainly employed in the Ministry of Rural and Land Development, the Land Consolidation and Development Authority, and the Sarawak Land Development Board. Some of these officials are members of the political parties in power; others may be supporters (Ngidang 2002: 169)

openly debated, for fear of being stigmatised. However, these important questions were discussed freely among perceived detractors.[25]

Not being able to clarify important issues breeds confusion and generates unnecessary division among perceived supporters and detractors within communities and across rival ones. More importantly, not being able to address questions that matter to communities places implementers at risk of not being able to plan efficiently. That is because the feedback they need in order to evaluate the social and economic sustainability of projects may be subject to two levels of filtration, as evident from my own research at Ulu Teru.[26]

For various reasons mentioned earlier, some of the Ulu Teru longhouses were regarded by officials as 'anti-government' and 'anti-development'. In the context of Konsep Baru, those considered 'anti' were merely concerned with the same sorts of issues that were harboured by the perceived 'supporters', who did not dare express their reservations openly for fear of being stigmatised. The difference was that the 'anti' group wanted to be better informed before they made decisions about Konsep Baru. These groups were consulted at the initial stages when officials visited Ulu Teru, but as a result of their daring to ask questions, they were labelled 'anti'. As the process unfolded, the 'anti' groups missed out on the series of dialogues held in Ulu Teru regarding Konsep Baru. Since officials were only comfortable dealing with their perceived supporters, the 'anti' groups were often not invited to these information sessions. To gain information, the perceived 'anti' group resorted to other means. Not losing faith in government, they visited the local government offices at Long Lama, Marudi or Miri. Local government offices are a mixed group of institutions with varying degrees of understanding about local issues. While some local officials may be politicised to the point of regarding the independent-minded groups as 'anti-government', others can be counted on for support, and rural longhouse dwellers learnt quickly about these differences. Longhouse people's creativity also led them to seek information from NGOs in Miri and Marudi.

Putting the different types of information together, several longhouses in Ulu Teru decided in 1998 that they would prefer not to be part of the plantation program under Konsep Baru. Since most implementers were not interacting with the more independent longhouses, they were not aware of changes in local sentiments (interviews at Ulu Teru, May 2000). In 1998, when longhouse residents resisted company bulldozers or stopped surveyors from working on their land, implementers were caught unaware. About 60 men and women blockaded bulldozers from what they regarded as 'trespassing' on their land. All were

---

[25] I have written about freely expressed concerns among perceived 'detractors' of Konsep Baru elsewhere (Majid Cooke 2002, 2003a).

[26] I worked at several longhouses in the middle Baram area, and intermittently at Ulu Teru, over a period of four months in April–May 2000 and August–September 2001. This section of the chapter draws on in-depth interviews with largely Iban groups from two longhouses of the area.

arrested and later released. This unfortunate episode was interpreted by many as an attempt to 'defend' their 'land and lifestyle'. Although many other longhouses in Ulu Teru were keen to accept Konsep Baru, the oil palm company pulled out because of unresolved 'sensitive issues', and the project had to be put on hold (see Ngidang 2002).

Conclusion: Development, the State and Localities

The process of persuasion that is taking place in Sarawak supports the possibility that the production of primary commodities such as palm oil is similar to raw material extraction (such as mining) because it is about the production and creation of marginal or frontier areas. For centuries, the frontier has been imagined as free for the taking, and opportunities abound through resource extraction and quick profits (Tsing 2000: 121). In 'frontier country', the culture is dedicated to the abolition of local land and resource rights as well as local commitment to landscapes. In line with contemporary analysis of state formation, this levelling process has been viewed here as an expansion of state spaces. The fundamental error which began in the Brooke period, and extended into the colonial era, is propagated during the 21st century using persuasion and legal codes that complete the unfinished business that may have accumulated during the colonial era. This chapter has shown that the oil palm story is not just about raising the standards of living of native communities — 'bringing them into the mainstream of development' so to speak — but is also about power, control and the expansion of state spaces. Such expansion by persuasion has been effective in some instances, but has been contested locally in other cases.

A culture of looking towards the government for 'development' is well entrenched in Sarawak (as it is in other parts of Malaysia). It is a culture that involves a trade-off. From the perspective of government and some community groups who are their supporters, the trade-off is worthwhile, the result being economic growth and the raising of living standards for many. From the perspective of community groups looking for more than economic growth, this trade-off may be too costly if what emerges from the exchange are measures that lead to a loss of their autonomy and a more dependent lifestyle.

With Konsep Baru the trade-off is once again being tested. The network of uncertainties surrounding the program, and the way they are addressed, shows up the lack of an avenue in the state system for the expression of non-economic rights. In a situation where the implementation machinery of a development program has been politicised, the feedback regarding Konsep Baru that enters the state system is one-sided and self-censored by party supporters, creating an inability to deal with non-economic development needs. Options for state creativity in dealing with social and political development are therefore closed. By extension, opportunities for understanding the real reasons for Konsep Baru's slow progress are closed as well.

Through strategies of control, state capacity to enforce the law against its citizens may enhance its ability to implement change in the economic sphere, but it is not necessarily the sign of a strong state in the social and political sphere. Reformists interested in decentralising state power are often able to notice and capitalise on this weakness by promoting change in the cultural sphere of economic development (Majid Cooke 2003b). By contrast, where the administrative machinery may be less politicised, as in the part of Sabah that Vaz (Chapter 7) worked in, feedback into the state system may come from a variety of sources and is less censored, contributing to a less rigid form of administration.

References

Appell, G., 1997. 'The History of Research on Traditional Land Tenure and Tree Ownership in Borneo.' *Borneo Research Bulletin* 28: 82–97.

Bian, B., 2000. 'Perspectives: Proposed Amendments to the Land Code (Land Code Amendment Bill 2000).' Unpublished report.

Bulan, R., 2000. 'Adong Bin Kuwau v. Kerajaan Negeri Johor: One Step Forward; Two Steps Backwards for Native Title?' Paper presented at the 'Borneo 2000' conference of the Borneo Research Council, Riverside Majestic Plaza Hotel, Kuching, 10–15 July.

Chua, T.K., 1992. 'Modernisation of Agriculture in Sarawak: Towards Achieving the Objectives of Vision 2020.' *Jurnal AZAM* 8(1): 86–121.

Cleary, M. and P. Eaton, 1996. *Tradition and Reform: Land Tenure and Rural Development in Southeast Asia*. Kuala Lumpur: Oxford University Press.

Cramb, R., 1989. 'Explaining Variations in Bornean Land Tenure: The Iban Case.' *Ethnology* 28(3): 277–300.

Dandot, W.B., 1992. 'Land Development Programme as a Development Strategy in Bidayuh Areas.' *Jurnal AZAM* 8(1): 141–161.

Dove, M.R., 1986. 'Peasant versus Government Perception and Use of the Environment: a Case-study of Banjarese Ecology and River Basin Development in South Kalimantan.' *Journal of Southeast Asian Studies* 17(1): 113–136.

———, 1999. 'Representations of the "Other" by Others: The Ethnographic Challenge Posed by Planters' Views of Peasants in Indonesia.' In T.M. Li (ed.), *Transforming the Indonesian Uplands*. Amsterdam: Harwood/Singapore: Institute of Southeast Asian Studies.

Embong, A.R., 2000. *Negara, Pasaran dan Pemodanan Malaysia [The State, the Market and Modernizing Malaysia]* . Banggi: Universiti Kebangsaan Malaysia.

Fong, J.C., 2000. 'Recent Legislative Changes on the Law Affecting Native Customary Land.' Paper presented at a seminar on the 'Land Code (Amendment) Ordinance, 2000', organised by the Sarawak Land and Survey Department, Kuching, 6 September.

Leach, M. and R. Mearns (eds), 1996. *The Lie of the Land: Challenging Received Wisdom on the African Environment*. Portsmouth (NH): Heinemann.

Majid Cooke, F., 1999. *The Challenge of Sustainable Forests: The Policy of Forest Resource Use in Malaysia, 1970–1995*. Sydney: Allen and Unwin/Honolulu: University of Hawaii Press.

———, 2002. 'Vulnerability, Control and Oil Palm in Sarawak: Globalisation and a New Era?' *Development and Change* 33(2): 189–211.

———, 2003a. 'Maps and Counter-Maps: Globalised Imaginings and Local Realities of Sarawak's Plantation Agriculture.' *Journal of Southeast Asian Studies* 34(2): 265–284.

———, 2003b. 'Non-government Organisations in Sarawak.' In M. Weiss and S. Hassan (eds), *Social Movements in Malaysia: From Moral Communities to NGOs*. London and New York: Routledge Curzon.

Ngidang, D., 2002. 'Contradictions in Land Development Schemes: The Case of Joint Ventures in Sarawak, Malaysia.' *Asia Pacific Viewpoint* 43(2): 157–180.

Peluso, N. and P. Vandergeest, 2001. 'Genealogies of the Political Forest and Customary Rights in Indonesia, Malaysia, and Thailand.' *Journal of Asian Studies* 60(3): 761–812.

Porter, A.F., 1967. *Land Administration in Sarawak: An Account of the Development of Land Administration in Sarawak from the Rule of Rajah James Brooke to the Present Time (1840–1967)*. Kuching: Government Printer.

Rangan, H., 1997. 'Property vs Control: The State and Forest Management in the Indian Himalaya.' *Development and Change* 28(1): 71–94.

Richards, A.J.N., 1961. *Sarawak Land Law and Adat: A Report*. Kuching: Government Printer.

Rousseau, J., 1987. 'Kayan Land Tenure.' *Borneo Research Bulletin* 19(1): 47–56.

Sarawak Ministry of Land Development, 1997. *Handbook on the New Concept of Development of Native Customary Rights (NCR) Land*. Kuching: Government Printer.

Sather, C., 1990. 'Trees and Tree Tenure in Paku Iban Solciety: The Management of Secondary Forest Resources in a Long-Established Iban Community.' *Borneo Review* 1(1): 16–40.

Scott, J.C., 1998. *Seeing Like a State: How Certain Schemes to Improve the Human Condition Have Failed.* New Haven (CT): Yale University Press.

Tsing, A., 2000. 'Inside the Economy of Appearances.' *Public Culture* 12(1): 115–144.

Vandergeest, P., 1996. 'Mapping Nature: Territorialization of Forest Rights in Thailand.' *Society and Natural Resources* 9: 159–175.

————— and N.L. Peluso, 1995. 'Territorialization and State Power in Thailand.' *Theory and Society* 24: 385–426.

Williamson, T., 2002. 'Incorporating a Malaysian Nation.' *Cultural Anthropology* 17(3): 401–430.

Chapter Three

Native Customary Land: The Trust as a Device for Land Development in Sarawak

Ramy Bulan

Introduction

The Sarawak government's strategy for economic growth through commercial development of agricultural land has resulted in vast areas of land being opened for large-scale plantations, including oil palm. In some places this has affected lands subject to 'native customary rights' (Sarawak Government 1997). When such rights are established over a tract of Interior Area Land, it becomes Native Customary Land. The latest type of development scheme, often dubbed Konsep Baru (New Concept), is one that uses the concept of fiduciary trust in the formation of joint ventures between native landowners, the government and large corporations.

This chapter examines native customary rights under existing legislation and the development strategy applied to Native Customary Land. It traces the chronology of past strategies and the culmination of those experiments in the joint venture concept. Are there any strengths in those strategies that may be built on or indeed be revisited? The evolution and rationale of a trust, and the nature of interests under a trust, are examined in the light of its suitability for the development of Native Customary Land in Sarawak. The duties of trustees and the fiduciary relationship are considered in order to ascertain the distribution of rights and possible remedies in the event of the trustees' breaches of duty. The chapter argues that, while the trust is a novel concept, the specifics of native customary rights in Sarawak may require further safeguards to be put in place to protect such rights.

Defining Native Customary Rights to Land

Sarawak has an anomalous and unique history as a British colony. A British protectorate in 1888, it was only annexed to British dominion in 1946 and became independent when it joined Malaysia in 1963. From 1841 to 1946[1] it was ruled by the Brooke family, whose members were themselves British subjects. This

---

[1] Sarawak was under Japanese occupation from 1942 to 1945.

historical legacy has shaped, and continues to influence, the development of the law and policies relating to native customary land.

Prior to James Brooke's arrival in Sarawak there was in existence a system of land tenure based on *adat* (native customary laws). That system remained virtually the same over the following century. Native customary rights to land consisted of rights to cultivate the land, rights to the produce of the jungle, hunting and fishing rights, rights to use the land for burial and ceremonial purposes, and rights of inheritance and transfer. According to native ideas, the clearing and cultivation of virgin land confers permanent rights on the original clearer (Geddes 1954; Freeman 1955; Richards 1961).

As the term implies, native customary rights may only be claimed by a native, or a person who has become identified with and has become subject to native personal law, and is therefore deemed to be a native.[2] 'Native' refers to the indigenous groups who inhabit the state, as listed in the schedule to the Sarawak *Interpretation Ordinance* and Article 161A, Clause 6 of the *Federal Constitution*. Despite the existence of numerous groups, the term 'Dayak' is colloquially used to refer to all the non-Muslim natives, differentiating them from the Malays, who by legal definition are Muslims (Bulan 1999; Hooker 2000). However, it is notable that the constitutional definition of natives in Sarawak includes the Malays. While the Malay-Melanau groups are coastal dwellers, the Dayaks are typically longhouse dwellers whose livelihood depends on the jungle and on swidden farming. Occupying the intermediate zones and the interior areas of Sarawak, their geographical locations and dependence on the land clearly determine the way that land administration affects them.

The Brookes did not interfere with the customary land rights of the Dayaks and Malays, allowing them a degree of self-governance. No scheme of alienation or land development was introduced except with respect to land where no rights or claims, whether documentary or otherwise, existed. There was a need to regulate the administration of land,[3] and at every phase, there was an awareness of the existence of native customary rights. As the authorities discovered, the regulation of customary tenure and land use touched on a social consciousness in which land has economic, social and religious significance (Porter 1967: 11).

---

[2] See Sections 8 and 9 of the *Sarawak Land Code* 1958 and Section 20 of the *Native Courts Ordinance* 1992.

[3] James Brooke's first attempt at codification of land tenure through the *Land Regulations* 1863 treated all land in the state as belonging to the government but only if it was 'unoccupied and waste lands'. Order VIII of 1920 consolidated and amended all preceding orders and defined state land to mean 'all lands which are not leased or granted or lawfully occupied by any person'. In 1931, Order L-2 redefined state land as 'all lands for which no document of title has been issued'. This was followed by Order L-7 of 1933, which required all lands to be registered on pain of nullity, and in effect marked the first introduction of the Torrens system in Sarawak, because it required an accurate cadastral survey as its basis, even though the government did not have the machinery to cope with a survey of the whole country.

After a number of regulatory orders, a memorandum on native land tenure was published by means of the Secretarial Circular No. 12 of 1939.[4]

Cultivated land and any land on which a fruit grove had been planted is heritable. Communities may also demarcate certain areas of primary jungle as *pulau* (reserved forest land) for communal use, within which rights over different resources may be established. Although judicial decisions have held these rights to have been lost upon personal abandonment, migration, or transfer, these losses must be seen in the light of the customary practices of each individual community.

Legislation on Land

English law was formally applied by the Brookes through *Order L-4 (Laws of Sarawak Ordinance)* 1928. This introduced English law subject to modifications by the Rajah, and was applicable to native customs and local conditions.

After the Brookes, the most significant period for Sarawak's land law was that which followed the cession of Sarawak to the British Crown in 1946. The Instrument of Cession transferred the rights of the Rajah, the Rajah in Council, and the State and Government of Sarawak in all lands to His Britannic Majesty 'but subject to existing private rights and native customary rights'. The *Application of Law Ordinance* 1949 provided for the reception (afresh) of English common law and doctrines of equity together with statutes of general application. These applied only 'so far as the circumstances of Sarawak and of its inhabitants permit and subject to such qualifications as local circumstances and native customs render necessary'.

One of the first pieces of legislation passed by the colonial government was the *Land (Classification) Ordinance* 1948. This instituted the system of land classification by which all land was divided into:

- Mixed Zone Land (land which may be held by any citizen without restriction);
- Native Area Land (land with a registered document of title but to be held by natives only);
- Native Communal Reserve (declared by Order of the Governor in Council for use by any native community, regulated by the customary law of the community);
- Reserved Land (reserved for public purposes);

---

[4] The Memorandum recognised the practice of rotational swidden agriculture, which by and large was and is widely practised among the natives. With slight variations, each community had a communal right to land, which was a right to occupation and exploitation in a general area within a territory without a clearly demarcated or rigid boundary. Individually, the original feller of a virgin jungle had an exclusive right to cultivate land which he had cleared. That land might be left to fallow as *temuda* (an Iban term), and then be recultivated after a number of years. Once it reverted to forest, it was available to the community for fishing, hunting or gathering of forest produce, but the 'pioneer' household retained the pre-emptive right over the *temuda* for recultivation.

- Interior Area Land (land that does not fall within the Mixed Zone); and
- Native Customary Land (land in which customary rights, whether communal or otherwise, have been created).

The effect of the classification was that the non-natives could acquire rights only in the Mixed Zone Lands. The natives were restricted in their dealings with non-natives, as well as among themselves, in line with the government policy of preventing the natives 'from impoverishing themselves by disposing lightly of their rights to others, whether alien or natives'. Native Customary Land was preserved wherever it was already created, irrespective of the zone. Any transfer or dealing contrary to the code was subject to a penalty (Porter 1967: 77).

A significant provision of the 1948 Ordinance was that natives were entitled to occupy Interior Area Land for the purpose of creating customary rights but they were to be licensees of the Crown. Since by definition a licensee holds land at the discretion of the owner, in one stroke that ordinance removed proprietary rights to land from people who for generations had occupied and depended on that land.

The reduction of native rights to a mere licence advanced the presumption that natives had only a usufructuary right with no kind of ownership, and underpinned the colonial 'tendency, operating often at times unconsciously, to render that title conceptually in terms which are appropriate only to systems which have grown up under English law'.[5] To deny the existence of a valid native perspective on land ownership, based on an elaborate system of rules and customs, was 'characteristic of the self-serving ethnocentricity upon which colonialism is based' (McNeil 1990: 92). The fact was that Sarawak was already inhabited by native groups who were not mere wanderers but were people in occupation of the land.

Amendments made through the *Land (Classification) (Amendment) Ordinance* 1955 precluded the creation of customary rights over Interior Area Land from 16 April 1955 unless a permit was obtained from the District officer. This continued to form the basis of the *Land Code* that came into force in January 1958, and remained an integral part of the land law system even after Sarawak joined Malaysia in 1963.[6] However, the issue of permits was effectively halted in 1964 by means of a government directive (Zainie 1994).

---

[5] 'Amodu Tijani v Secretary of State, Nigeria', *Appeal Cases* 1921(2): 399.
[6] At the time of joining the Federation of Malaysia, the natives expressed the need to safeguard their customary rights to land. In 1962, a Commission of Inquiry instituted under Lord Cobbold of England recommended, among other things, that 'land, agriculture, forestry and native customs and usages' should be under the control of the state governments. The Federal Constitution thus bestowed the state governments with the jurisdiction in those matters under Article 64, Schedule 9, and the legal status quo in relation to the customary lands of Sabah and Sarawak has been allowed to remain. The federal government has limited powers to pass laws on land solely with the purpose of unifying the law.

The *Sarawak Land Code* 1958

The *Sarawak Land Code* 1958 is based on a Torrens registration system which only recognises registered interests in land. The person claiming ownership or interest must have a document of title in the form of a grant, lease or other document as evidence of title or interests. There is, however, a provision for the creation of Native Customary Land under Section 5(2) which is limited to six specific methods; namely:

- the felling of virgin jungle and the occupation of the land thereby cleared;
- the planting of land with fruits;
- the occupation of cultivated land;
- the use of land for a burial ground or shrine;
- the use of land for rights of way; and
- by any lawful method (deleted in 2000).

Numerous amendments have been made to the *Land Code*. For instance, in 1994 amendments were passed to empower the minister in charge of land matters to extinguish native customary rights to land. In 1996, the onus was placed on a native claimant to prove that he has customary rights to any land against a presumption that the land belongs to the State. In 1998, to pave the way for extinguishment or compulsory acquisition of land, the mechanisms for assessment and payment of compensation were put in place.

The most comprehensive set of amendments were those set out in the *Land Code (Amendment) Ordinance* 2000. This included a definition of 'native rights' which was curiously missing in earlier legislation. Section 7A(1) streamlines 'native rights' into three categories; namely:

- rights lawfully created pursuant to Section 5(1) or (2);
- rights and privileges over any Native Communal Reserve under Section 6(1); and
- rights within a kampung reserve (Section 7).

The 2000 amendment harmonised the processes and procedures relating to Native Customary Land with those relating to other types of alienated land in respect of the resumption of land and the adjudication of payable compensation for termination of rights. It also provided for the creation of a Registry of Native Rights. Finally (and notably), the amendment deleted 'any lawful methods' under Section 5(2)(f), for what Fong (2000) described as the sake of legal certainty and clarity.

Some lawyers have argued that the implicit intention of the legislature in 1958 would have been to make a provision for certain customs and practices not covered by the *Land Code* (Bian 2000), but which were observed by different groups under their customary laws. The practice of customary land tenure certainly did not cease in 1958 and, as Bian argues, some lands had been acquired through barter exchange or 'sale' within communities, or as marriage dowries, which were subsumed under the 'other lawful methods'. Given the inherent flexibility of *adat* (Cramb 1989; Sather 1990), and its ability to adapt to demographic and economic changes, Bian's argument is reasonable.

The restricted concept of native customary rights under Section 5 made it difficult to assert rights under the *Land Code* after 1958 (Bulan 2000). The line of restriction is not a new phenomenon (Majid Cooke 2002). As Porter commented on the inception of the code, it 'virtually prohibit[s] the creation of new customary rights' and the 'extremely detailed and rigid' provisions 'dictated government policy' (Porter 1967: 83, 99).[7] Fong (2000) argues that the intention of the subsequent amendments was to restrict the methods of creating native customary rights to those stipulated under Section 5.

It is significant, therefore, that in a recent court case, Ian Chin recognised the existence of the Iban *pemakai menoa* — the area from which its members 'eat' (*makai*) — within which are found their *temuda* (secondary forest) and the *pulau galau* (land reserved for communal use).[8] The concept of *pemakai menoa* goes beyond mere agricultural use and extends to hunting, fishing and living off the produce of the jungle. Ian Chin held that those customary rights had not been expressly abolished by earlier orders or other legislation.

The Court of Appeal overturned the High Court decision on 9 July 2005,[9] holding that there was insufficient proof of occupation by the (Iban) respondents in the disputed area, although they had satisfied the test for native customary rights in the adjacent area. Nonetheless, the Court of Appeal did not disturb the High Court's finding that the Iban concept of *pemakai menoa* exists. This is a milestone for native customary rights in Sarawak

The Court of Appeal endorsed the exposition of the law by the learned judge of the High Court when he argued that the common law respects the pre-existence of rights under native laws or customs and that these rights do not owe their existence to statutes. Legislation is only relevant to determine how many of those native customary rights have been extinguished. It affirmed that the *Land Code* does not abrogate native customary rights that existed before the passing of that legislation, but held that natives are no longer able to claim new territory without a permit from the Superintendent of Lands and Surveys under Section 10 of the code. It also agreed with the High Court that the rights held under a licence 'cannot be terminable at will', for they can only be extinguished in accordance with laws subject to payment of compensation. Any discussion of

---

[7] See Sarawak Government 1959, Paragraph 27; *Land Code*, Section 10(3) and (4); Adam 1998: 217.

[8] 'Nor Anak Nyawai and Ors v Borneo Pulp Plantation Sdn Bhd and Ors', *Current Law Journal* 2001(2): 769.

[9] 'Borneo Pulp wins appeal case on NCR land' (*Sarawak Tribune*, 9 July 2005).

the development of native customary rights must therefore bear in mind that, despite the provision of Section 5, the native concept of land is broader than the restrictive statutory provisions.

As the state seeks to accelerate land development under its broader 'politics of development' (Jitab and Ritchie 1991), the medium that is felt best suited to bring 'development and progress' to the natives is estate development. This involves lands which some native groups claim to be their communal lands.

Agricultural Policies and Land Development Schemes

To encourage native smallholders to participate in commercial land development, a series of land development schemes were undertaken from the 1960s to the 1980s. These have been documented by many writers such as Hong (1987), King (1988), Cleary and Eaton (1996), Ngidang (1998) and Majid Cooke (2002).

From 1964 to 1974, land resettlement schemes modeled after the integrated style of development adopted by the Federal Land Development Authority (FELDA) of Peninsular Malaysia were introduced and implemented — initially through the Agriculture Department and later (1972–80) through the Sarawak Land Development Board (SLDB). This involved clearing of new land and relocation of natives into resettlement schemes dedicated to the planting of cash crops (Ngidang 1997). Unlike the FELDA schemes, where landless workers were settled on state land, participants in Sarawak were relocated to areas where the local communities were established traditional landowners. The farmers were given loans that they had to repay out of incomes which were crucially dependent on the fluctuations of world commodity prices, and as a result, most were unable to make the repayments. The schemes also lacked the pool of workers and expertise required for their successful implementation (King 1988: 280) and all were eventually abandoned due to management problems (Ngidang 2001).

In 1976, the Sarawak Land Consolidation and Rehabilitation Authority (SALCRA) was established with the object of developing agricultural land *in situ* (Hong 1987; King 1988: 283). This was a joint venture between the SALCRA and native farmers in which the participating households retained their ownership (Munan 1980; Humen 1981: 95–106). Subject to payment of costs by the owner, large consolidated blocks of land have been planted with cash crops. The SALCRA's function includes consolidation and rehabilitation of land, and provision of advisers and training facilities in various aspects of farming and land management. When it appears that the participants have acquired the know-how to manage the schemes, the estate should be divided among the households, thus enabling them to obtain a demarcated piece of land to which they will be given a grant in perpetuity. Although there has not been any rationalisation and distribution exercise yet,[10] the eventual obtaining of titles for landowners through their participation appears to be an ideal solution to the problem of modernising agriculture in many native areas. Substantial alienation of land to non-native private companies with commercial interests has been avoided. To some extent, rural–urban migration has also been arrested. However, the success of the scheme is dependent on continued government funding.

Parallel to the SALCRA, the Land Consolidation and Development Authority (LCDA) was set up in 1981 to promote the development of both agricultural and non-agricultural projects. The LCDA has powers to acquire both state-controlled land and Native Customary Land for private estate development. It has powers to act as an intermediary between landowners and corporations so that private investors can be invited to participate in land development subject to allocation of shares in the relevant companies. The *Land Code* was amended in 1988 and 1990 to allow corporations, including foreign companies, to purchase land, including Native Customary Land, for this kind of development.

The formation of the LCDA was a further step in government involvement in large-scale land development as it became an agency and a conduit to 'harness private capital for developing the land as estates' (Sarawak Government 1997). This paved the way for the introduction of the joint venture company (JVC).

The New Model: Joint Venture Companies

The concept of the joint venture is premised on the assumption that Native Customary Land, which is now unorganised and fragmented, can be turned into an economic asset through the creation of a Native Customary Land Bank. Once pooled, it is assumed that large-scale plantation development and optimum returns can be realised. It is also assumed that large areas of Native Customary Land are attractive and viable for private investment.

As a prerequisite, there should be contiguous blocks of land of not less than 5000 hectares, which may cover land spanning the territorial domain of several longhouse communities. To date, the SLDB and the LCDA have been appointed as managing agents. Every landowner has to sign a trust deed to assign to the government agency all their respective rights, interests, shares and estate in the land. The government agency will then enter into the joint venture with the private corporation. When an area is marked for commercial development, the Minister may declare that area of land as a Development Area under Section 11 of the *Land Consolidation and Development Authority Ordinance* 1981, and the land will be classified as Native Area Land under Section 9(c) of the *Land Code*.[11]

---

[10] Personal communication, R.J. Ridu, former Speaker of Negeri Council, August 1999.

[11] Where a non-native has been issued with any permit relating to Native Area Land or Native Customary Land, the SLDB shall apply to the *Majlis Mesyuarat Kerajaan Negeri*, or the State Secretary to whom powers have been delegated, for a special direction that the company be deemed a native under Section 9(1)(d) of the *Land Code*.

A perimeter survey using the global positioning system is carried out by the JVC to determine the size of the area. One title will be issued to the JVC for a period of 60 years (two plantation cycles) for an agreed value. The owners have to agree amongst themselves to determine the approximate sizes of their landholdings, and their names are then to be listed in the appendix of the trust deed. Under Section 18 of the *Land Code*,[12] the Superintendent of Lands and Survey may issue a lease over any land within a Development Area[13] for a term of not more than 60 years to a body corporate approved by the Minister. All adjoining land may be amalgamated.

In consideration for the use (or lease) of Native Customary Land, the JVC will issue to the trustee shares in the JVC credited as fully paid, equivalent to 60 per cent of the value of the said land, representing 30 per cent of the issued and paid up capital of the JVC. The value has been pegged at RM1200.00 per hectare. The JVC will pay to the trustee the equivalent balance of 40 per cent of the said land value. Out of that sum, 30 per cent will be invested in government-approved unit trusts and 10 per cent will be paid to the landowners.

In terms of equity ratio, the trustee will pay cash for 10 per cent of the issued share capital and the private company developer will pay cash for its 60 per cent share, while the landowners' equity of 30 per cent in the JVC will be paid through the land value. The said land may only be used for agricultural purposes, and the JVC cannot deal with or charge the land as security for loans to implement the project without the prior approval of the Minister.

Upon expiry of the term of the title, the customary landowners shall decide either to renew the title in favour of the JVC or request for the land to be alienated to themselves or to another company or another entity nominated by them in writing. In the event that the customary landowners are desirous to have the land subdivided and alienated to them individually, the JVC is empowered to undertake a survey of the land and to determine the most equitable and fair manner of subdivision, having due regard to the extent of each of the landowners' interest in the land. This is the stage at which the distribution and allotment of shares may be problematic because of uncertainty about the size of people's shares in the land.

Two pioneer schemes have been developed as pilot projects in the Baram and Kanowit districts (Ngidang 1997: 75; Songan and Sindang 2000: 251) with varying responses from the participants. Many people participated in the projects without a full understanding of what such alien concepts as the trust, a joint venture,

---

[12] By means of Section 6 of the *Land Code (Amendment) Ordinance* 1997.
[13] By Section 11 of the *Land Consolidation and Development Authority Ordinance* 1981 and the Schedule to the *Land Development Board Ordinance* 1972.

or shares in a company entailed.[14] This could give rise to the question of whether there has been effective consultation and informed consent on the part of the participants.

The next section discusses traditional rules relating to trusts and trustees, considers how those principles are applied in the JVC arrangement for development of Native Customary Land, and looks at possible remedies for landowners in the event of any breach.

The Trust and Protection of Property

The trust has tremendous utility because it is flexible and easy to create. It is usually set up for the purpose of 'the management of wealth', where property may be put on trust for an individual, an infant, a person of unsound mind or a group (Hayton 1998).

The modern trust evolved as a response of equity to the shortcomings and the rigid formalities of the common law. The trust was originally used to protect landowners who had transferred their land to another on the understanding that the transferee was to hold and administer the affairs relating to the land for the benefit of the transferor's family.

There would be no problem where the transferee kept his word, but when the transferee broke his promise, misused or administered the land for his own benefit, the question of rights and remedies would arise. Common law only recognises the ownership of the legal owner. In case of a breach, there would be no legal redress for the transferor and his family. The concept of the trust was a developed as a way of requiring the friend to fulfill his promise, on the basis that it was unconscionable for him to claim the land for himself. In other words, equity imposed a *trust* on the transferee, called the trustee, to hold the property for the benefit of the beneficiaries.

The reliance placed on the transferee to deal with the property for the benefit of the beneficiary gave rise to a relationship of confidence or a *fiduciary* relationship. The transferee could not deal with the property in a way that would conflict with the interest of the transferor or the equitable owner.

Today the trust has become a valuable device in commercial and financial dealings where the fundamental principles of equity that were originally formulated apply as much to commercial trusts as they do to the traditional trusts.

[14] The Baram project commenced in February 1997, involving 550 households from about 14 longhouses in a joint venture between Perlis Plantation Berhad and the SLDB, but the company reportedly withdrew from the project in early 2002. In Kanowit, 17 Iban longhouse communities are involved in a joint venture project between Kuala Sidim Bhd (a subsidiary of Boustead-Estate Bhd) and the LCDA acting as the trustee for the native customary landowners. A report of the Sarawak Development Institute found that 68.7 per cent of landowners in this area supported the JVC despite a low level of understanding, while 46.35 per cent did not understand the concept at all.

The Nature of the Trust

The significant feature of the trust is the dual ownership of the trustee (legal ownership) and the beneficiary (equitable ownership). There are four essential elements of a trust under ordinary principles of law:

1. There must be a trustee or somebody who holds the trust property.
2. There must be property, whether land or money, that is capable of being held on trust.
3. There must be an ascertained beneficiary or beneficiaries who could enforce their rights.
4. The trustee is under a personal (equitable) obligation to deal with the property for the benefit of the beneficiaries.

In 1840, Lord Langdale laid down three certainties in the creation of a trust, namely: the person establishing the trust (the settlor) must demonstrate a clear intention to create a trust; the subject matter (the beneficial interest) is clearly identified; and the beneficiaries as well as the quantum of entitlement must be certain.

Uncertainty of intention will cause the trust to fail, and the person on whom the gift is bestowed will take the gift absolutely unhampered by the trust. If the subject matter is not certain, no trust is created. It may be, however, that the property itself is certain but the beneficial shares are not. Unless the trustees have discretion to determine the amounts, the trust will fail and the property springs back to the settlor on a resulting trust (Martin 1997: 93). The beneficiaries of the trust must also be ascertainable, otherwise a trust would fail for uncertainty and the property reverts to the settlor (Hayton 1998: 82).

Apart from these three certainties, no rigid formalities are required. In Peninsular Malaysia, a trust concerning any property, including land and equitable interest in land, need not be in writing provided the words used in the transaction show a clear and unequivocal intention to create a trust.[15] In Sarawak, however, a declaration of trust in respect of any interest in land, whether legal or equitable, must be in writing signed by a person who is able to declare the trust or by his will.[16]

The Trust and Native Customary Land Development

This 'new model' JVC is a type of development trust which is a 'facilitative commercial trust' (Bryan 2001). Creating a trust circumvents the requirements

[15] 'Wan Naimah v Wan Mohamed Nawawi', *Malayan Law Journal* 1974(1): 41.
[16] 'Lee Phek Choo v Ang Guan Yau and Anor', *Malayan Law Journal* 1975(2): 146. The rationale for this, as explained by Chief Justice Lee Hun Hoe, is that Section 2 of the *Application of Laws Ordinance* and Section 3(1) of the *Civil Law Act* 1956 import into Sarawak and Sabah not only the rules of common law and equity but statutes of general application as well, hence Section 9 of the English *Law of Property Act* 1925, which requires that land transactions be in writing.

for a person or persons to be a party to the contract in order to enforce it. A third party cannot enforce the contract but he may enforce a trust even though he was not party to it. The beneficiaries include persons whose names appear in the appendix of the trust deed, their respective heirs, successors in titles, executors, administrators, personal representatives, trustees and any other person claiming title or interest in the name or on behalf of the native customary owners.

The trust also does away with the need to get into a partnership that will require the parties to contribute equally in order to share equally in the profits (Ladbury 1987). Most native landowners do not have the financial means to develop the land, so vesting the land in trustees is arguably one of the most appropriate mechanisms that can be used. Be that as it may, the intrinsic nature of native customary rights could give rise to problems peculiar to this kind of trust.

The JVC and the Nature of the Beneficiaries' Interests

The terms of the trust deed presume that the native customary owners have acquired the rights through one of the means prescribed under Sections 5(2), 7A, 7B or 7C, or had obtained a permit under Section 10, of the *Land Code* , or that there is evidence or records kept by the Land Office pertaining to the land, so that a registrable document of title may be issued in favour of the company.

This arrangement is different from some property development ventures which are financed through the marketing of shares in land trusts where the shares have clear proportions. In this case, while the beneficiaries may be entitled to the land as set out in the appendix of the trust deed, their respective interests, rights, and shares are undivided. With one master title, the owners cannot apply for subdivision for as long as the company is the registered proprietor.

Applying the traditional requirements of certainty, there is no exhaustive listing of all beneficiaries entitled under customary law. The question in this case is: could the trust be challenged as void for uncertainty of objects? If so, who is responsible to ensure that the land reverts to the owners? And finally, what are the powers of the trustees?

General Powers and Duties of Trustees

The trustee's powers are provided for by the trust deed, although general statutory powers are also provided by the *Trustee Act* 1949. The powers of a trustee are facilitative, enabling a trustee to act in a certain way but leaving the discretion to him as to whether to so act. Duties, on the other hand, are imperative. They compel or prohibit a trustee from acting in a certain way, failing which he may be liable for breach of duty.

The general powers of trustees under the *Trustee Act* include the powers to compound liabilities, to settle claims and to give receipts, to fix the value of trust property, to concur with co-owners of land in disposing of trust property, and to insure trust property.

The trustee cannot put himself in a position where there is a conflict of interest, nor can he profit from his position without authorisation by the trust deed or consent of the beneficiaries. It is his duty to administer the trust honestly and impartially for the benefit of the beneficiaries, to account to the beneficiaries and to distribute the income to those entitled to it.

A breach of duty may result in a claim by the beneficiaries. Any loss caused by the trustee or trustees wrongfully disposing of the assets or any diminution in the value of the trust fund may have to be borne by the trustees. The same liability may be imposed on a trust corporation, although the standard of care and business prudence expected of a trust corporation is higher than that of an ordinary trustee, particularly where it holds itself out as capable of providing certain expertise which cannot be provided by an ordinary prudent man. The reasonable standard of care is one for the courts to decide based on the facts of the case.

Underlying these powers and duties is the fiduciary obligation of the trustee to the beneficiary.

The Fiduciary Relationship and its Ramifications

The word 'fiduciary' comes from the Latin *fiducia* meaning 'trust'. Inherent in the nature of the fiduciary relationship is one party's position of disadvantage or vulnerability which causes him to place reliance upon another and requires the protection of equity in acting upon the conscience of that other. It is important to determine whether a fiduciary relationship exists and, if so, whether any remedy is available in case of any breach of that fiduciary obligation.

The relationship between a trustee and the beneficiaries has been called the 'archetypal' fiduciary relationship.[17] It is an established principle that the trustee must not use his position to make a gain for himself. This has been extended to apply generally to all cases where one person stands in a position of influence over another, enabling the court to intervene in circumstances where the person occupying a position of trust or confidence took improper advantage of that position. The question is: would these principles of fiduciary duty apply to a government and its agencies?

Dal Pont and Chalmers (1996: 118) argue that the government, like a trustee, is concerned with the control and distribution of wealth. Having been sourced from the people, the exercise of a government's power to affect the interests of

---

[17] The categories are not closed but this may include the relationship between directors and company, principal and agent, solicitor and client, guardian and ward, parent and child, or partners in a joint venture.

its people is subject to an obligation to deal with this wealth for the benefit of its people. In this respect, the people can be characterised as 'beneficiaries' of the trust established by the conferral of their authority on the government to act on its behalf (Finn 1994: 45). The fiduciary duty that binds the Crown is similar to the duty that a constructive trustee owes to a beneficiary, which entails a duty not to compromise the beneficiary's interest in transactions with third parties.

The highest courts in the United States, Canada, New Zealand and, to some extent, Australia have recognised the existence of a fiduciary relationship between the government and aboriginal persons. The issue of fiduciary obligation towards aboriginal people has also arisen in Malaysia. In one recent case,[18] the federal and state governments were both said to have owed a fiduciary duty to the Orang Asli (aborigines) of Peninsula Malaysia to protect them from unscrupulous exploitation and to safeguard their tribal organisation and way of life. That duty emanates from Article 8(5) of the *Federal Constitution*. This was affirmed by the Court of Appeal in 2005.

In the case of natives in Sarawak, Article 153 of the *Federal Constitution* also imposes a fiduciary obligation on the Yang di-Pertuan Agong (the King) to protect the interests of the natives of Sarawak and Sabah. Further preferential treatment as regards alienation of land by the state is contained in Article 161A(5), while protection of native law and custom is also enshrined under Article 150(6A), Clause 5.[19] Clearly, there is legal recognition that natives are especially vulnerable to the power of government, and this justifies their preferential treatment. For natives in Sarawak, this is a reflection of the Brooke government's belief that Sarawak 'is the heritage' of its people and that land is their 'lifeblood'. In the 'Nine Cardinal Principles of the Rule of the English Rajahs', the government held itself as 'trustee' of the people and policies for protection of native interests against outside exploitation were put in place.[20]

The state's fiduciary duty also arises because of the inalienability of the property. The state's power to impair native customary rights by way of alienation, and the fact that such rights are inalienable except to another native or by surrender to the state, gives rise to a fiduciary obligation on the state. The fiduciary obligation protects those rights so that they cannot be terminated without involving, informing, consulting and negotiating with the customary right holders in good faith, minimising the impact and detriment on the affected parties. It is imperative for the government to deal with the property surrendered to it with utmost good faith.

[18] 'Sagong Tasi and Ors v The Government of Selangor and Anor', *Malayan Law Journal* 2002(2): 591.
[19] In a period of emergency, Parliament may legislate on any matter, but that power shall not extend to any matter of Islamic law or the custom of the Malays or with respect to any matter of native law or custom in the states of Sabah or Sarawak notwithstanding anything in the Constitution.
[20] See the *Constitution Order No C-21 (Constitution)* 1941.

This means that, when native customary landowners surrender their rights to the LCDA as trustees, there is a clear fiduciary duty to protect the rights of the vulnerable right holders. A government agency that takes on the duties of a trustee under a commercial arrangement becomes a 'trustees twice over' (Finn 1992: 243), particularly where the vulnerable landowners depend on it to negotiate the best terms on their behalf (Lehane 1985: 98).

In the present JVC model, the relationship between the corporate developer and government agency (trustee) is contractual. Does a fiduciary relationship exist between them? It is suggested that the mutual confidence between the JVC and the LCDA (or its agents), in appropriate circumstances, does not exclude the possibility of a fiduciary relationship.

The Malaysian Federal Court has already held that the relationship between parties in a joint venture agreement is a fiduciary relationship.[21] Thus, if a right is not sustainable in breach of contract, there may be an avenue in equity where there is a breach of the fiduciary obligation.[22]

Breach of Trust and Remedies of Beneficiaries

What remedies are available to beneficiaries should there be any unauthorised act or in case of a breach?

At the core of the trust concept is also a right of the beneficiaries to make the trustees accountable to the trust and to ensure that they act within the terms of the trust deed. The remedy for the breach of a fiduciary duty includes declaration of rights or a claim in damages and compensation.

Beneficiaries have a right to have the trust property invested in a way that will keep a balance between them. They have a 'policing' right, to see the trust accounts from time to time, and to require the trustees to make good any breach of trust. While trustees are not bound to give reasons in exercising their discretion, the absence of reasons could create a presumptive case that a trustee's discretion has been miscarried or was not exercised upon real, sound and genuine consideration. Beneficiaries may also apply for an injunction to restrain a fiduciary from acting in a way that is detrimental to the trust.

The issue takes on a different angle where the trustee is a government agency. Section 29(1) and (2) of the *Government Proceedings Ordinance* 1956 debars an injunction being granted against a government or an officer of the state. An order for the preservation of property may be made if the plaintiff can show that irreparable damage not compensatable by damages would be caused. Despite

[21] 'Sri Alam Sdn Bhd v Newacres Sdn Bhd', *All Malaysia Reports* 1996(3): 3293.
[22] 'Tengku Abdullah ibni Sultan Abu Bakar v Mohd Latiff bin Shah Mohd', *Malayan Law Journal* 1996(2): 265. Gopal Sri Ram (Court of Appeal) said that, depending on the commercial morality, courts in a particular jurisdiction may choose to impose a fiduciary obligation on parties to a transaction having regard to the cultural background and circumstances of the society in which they function.

the nomenclature, if the effect is the same as that of an injunction, it will not be granted. Thus, while it is open for claimants to take legal action to prove their claims, very few natives have the means to sustain such actions.

Questions of Proof and Reversion of Land

A fundamental aspect of the JVC is that native customary 'owners' become joint venture partners without having to provide financial capital. This means that 'their equity in the joint ventures would be based on the area of their land; and the irresistible part of it all is that their land would be returned to them when the government has no more use for it' (Jitab with Ritchie 1991: 66). To what extent can the beneficiaries be assured of the reversion of the Native Customary Land? The issue is not that 'the government will not cheat its own people'; rather, the problem lies in the discharging of the onerous burden of proof that is on the claimant.

Since Native Customary Lands are not individually surveyed, there are latent uncertainties in terms of the specific shares in the land. At the expiration of 60 years, persons who have surrendered their rights may no longer be alive. This could cause problems for the successors unless they can work out a clear system of partition and inheritance of the land. If the native claimants are not able to settle their claims among themselves, there is a possibility and danger of them losing their rights to the legal owner who has a registered (master) title to the land.

The new Section 7A of the *Land Code* provides for registration of Native Customary Land but does not provide indefeasibility of title. In Fong's words, it is treated merely as an acknowledgement of a claim to the land until the contrary is proved,[23] a certification to a right, and not a 'proprietary right in land'. The onus of proving an interest remains on a native claimant.[24] The problem reverts to the question of the restrictive provisions under Section 5 and the clash between statute and native concepts of land.[25]

The commonly deployed method of determining the existence of native customary rights over a parcel of land is aerial photographs taken prior to 1 January 1958. However, these may not be available, and the claimant then has to show alternative physical evidence of occupation before 1958, or else show records of permits, which are virtually non-existent. Thus, upon amalgamation

---

[23] The *Land Code (Amendment) Ordinance* 1996 provides that 'whenever any dispute shall arise as to whether any native customary right exists or subsists over any state land, it shall be presumed until the contrary is proved, that such state land is free and not unencumbered by such rights.'

[24] *Land Code Amendment Ordinance* 2000, Section 7(A)(3)(b).

[25] This is a stark contrast to Section 66 of the Sabah *Land Ordinance*, where a native who establishes customary tenure and who has cultivated unalienated land for three years may apply for that land to be registered as native land and thereby acquires indefeasible interest, alleviating much of the anguish over uncertainty of title.

of all the contiguous lands by the Director of Lands and Surveys, land that is vested in the trustee becomes the legal property of the agency with no compensation paid to the claimant. With no payment of compensation at the point of amalgamation, is the amalgamation tantamount to summary taking of land without compensation?

One possible way to avoid this problem may be to survey the land and grant individual titles to the owners at the point of their joining the scheme. This would ensure that persons who join the scheme know their specific share and are able to stake a claim at the expiration of the 60-year provisional lease period. Before such a survey can be carried out, such rights must be fully investigated, demarcated and recorded before titles can be issued to replace the customary tenure (Goh 1969: 4). It has been argued that a full-scale statewide registration of native interests over land would be a time-consuming, tedious and costly operation (Fong 2000). However, in specific projects such as this, the advantages of a proper survey being done prior to implementation cannot be understated.

A prior grant of title to claimants would best serve the interest of the vulnerable owners and the sense of security would be an incentive for participation. It would go a long way in improving the implementation of Native Customary Land development (Songan and Sindang 2000: 251). With the passing of the *Land Surveyors Ordinance* 2002, the combined effect of Sections 20 and 23 entail that a person who is not a licensed surveyor cannot make, authorise or sign any cadastral map. Since map making by the communities themselves could be an offence, it is imperative that the authorities take steps to survey the land for the natives.

Concluding Remarks

The caution in commercial joint ventures is that, all too often, when there is no more money in the venture, it is easy for parties to forget their contractual obligations and the vulnerable parties often suffer. As an active sponsor of these schemes, it is all the more pertinent for the government to provide some kind of a guarantee that Native Customary Land will revert to the owners at the end of the venture. Similarly, in the event that a JVC withdraws without completing its job, is there some form of a guarantee the native customary owners will be adequately compensated?

Arguably the government's fiduciary obligation may be said to go beyond a mere commercial arrangement to become 'trustees twice over', based as it is on the customary owners' trust and confidence in the government. Perhaps there is scope here for application of Lord Browne-Wilkinson's (1995) caution 'that equity principles must follow developments in commercial law for commercial expediency, but such application has to be both thoughtful and sensitive'. What has been developed as an instrument to defeat unconscionable conduct should not ironically become the very instrument that defeats the rights of those that it purports to protect.

References

Adam, F.J., 1998. 'Customary Land Rights under the Sarawak Land Code.' *Journal of Malaysian Comparative Law* 25: 217–231.

Appell, G.N., 'The History of Research on Traditional Land Tenure and Tree Ownership in Borneo.' *Borneo Research Bulletin* 28: 82–97.

Bian, B., 2000. 'Perspectives: Proposed Amendments to the Land Code (Land Code Amendment Bill 2000).' Unpublished report.

Browne-Wilkinson, Lord, 1995. 'Equity and Commercial Law: Do They Mix?' Tenth Sultan Azlan Shah Law Lecture, Kuala Lumpur, 22 December.

Bryan, M., 2001. 'Reflections on Commercial Applications of the Trust.' Paper presented at a conference on 'Key Developments in Corporate Law and Equity', Monash University, 16 March.

Bulan, R., 1999. 'Native Status and the Law.' In M.A. Wu (ed.), *Public Law in Contemporary Malaysia*. Petaling Jaya: Longman Malaysia.

———, 2000. 'Indigenous Peoples and Property Rights to Land: A Conceptual Framework.' Paper presented at a workshop on 'Customary Land Rights: Recent Developments', Faculty of Law, University of Malaya, 18–19 February.

———, 2001. 'Native Title as a Proprietary Right Under the Constitution in Peninsula Malaysia: A Step in the Right Direction?' *Asia Pacific Law Review* 9(1): 83–101.

Cleary, M. and P. Eaton, 1996. *Tradition and Reform: Land Tenure and Rural Development in Southeast Asia*. Kuala Lumpur: Oxford University Press.

Cramb, R., 1989. 'Explaining Variations in Borneo Land Tenure: The Iban Case.' *Ethnology* 28: 277–300.

———, 1990. 'The Role of Smallholder Agriculture in the Development of Sarawak 1963–1988.' *Jurnal Azam* 6: 103–123.

Dal Pont, G.E. and D.R.C. Chalmers, 1996. *Equity and Trust in Australia and New Zealand*. North Ryde (NSW): Law Book Company Information Services.

Finn, P.D., 1992. 'Integrity in Government.' *Public Law Review* 3: 243–257.

———, 1994. 'The Abuse of Public Power in Australia: Making Our Governors Our Servants.' *Public Law Review* 5: 43–45.

Fong, J.C., 2000. 'Recent Legislative Changes in the Law Affecting Native Customary Land.' Paper presented at a workshop on 'The *Land Code (Amendment) Ordinance* 2000', Kuching, 6 September.

Freeman, J.D., 1955. *Report on the Iban of Sarawak*. Kuching: Government Printer.

Geddes, W.R., 1954. *The Land Dayaks of Sarawak*. London: HM Stationery Office.

Goh, M.T., 1969. *Brief on Sarawak Land Tenure System*. Kuching: Sarawak Lands and Surveys Department.

Hayton, J.D., 1998. *The Law of Trusts: Fundamental Principles of Law* (3rd edition). London: Sweet and Maxwell.

Hong, E., 1987. *The Natives of Sarawak: Survival in Borneo's Vanishing Forest*. Pulau Pinang: Institut Masyarakat.

Hooker, M.B., 1979. 'Native Law in Sabah and Sarawak.' *Malayan Law Journal* 1979(2): 30–38.

———, 1999. 'A Note on Native Land Tenure in Sarawak.' *Borneo Research Bulletin* 30: 28–40.

Humen, G.G., 1981. *Native Land Tenure Protection in Sarawak*. Kuala Lumpur: University of Malaya (LL.B. Hons. thesis).

Jegak, U., 2001. 'Awareness, Perception and Acceptance of the New Concept for Development of NCR in Dijih/Sebaking, Mukah.' *Sarawak Development Journal* 2(2): 59–78.

Jitab, K. with J. Ritchie, 1991. *Sarawak Awakens: Taib Mahmud's Politics of Development*. Selangor Darul Ehsan: Pelanduk Publications.

King, V.T., 1988. 'Models and Realities: Malaysian National Planning and East Malaysian Development Problems.' *Modern Asian Studies* 22: 263–298.

Ladbury, R.A., 1987. 'Commentary.' In P.D. Finn (ed.), *Equity and Commercial Relationships*. Sydney: Law Book Company.

Lehane, J.R.F., 1985. 'Fiduciaries in a Commercial Context.' In P.D. Finn (ed.), *Essays in Equity* . Sydney: Law Book Company.

Leigh, M.B. (ed.), *Environment, Conservation and Land: Proceedings of the Sixth Biennial Borneo Research Conference*. Kuching: UNIMAS.

Majid Cooke, F., 2002. 'Vulnerability, Control and Oil Palm in Sarawak: Globalisation and a New Era?' *Development and Change* 33: 189–211.

Martin, J.E., 1997. *Hanbury and Martin Modern Equity* (15th edition). London: Sweet and Maxwell.

McNeil, 1990. 'A Question of Title: Has the Common Law been Misapplied to Dispossess the Aboriginals.' *Monash Law Review* 16: 91–110.

Munan, S., 1980. 'Sarawak Land Consolidation and Rehabilitation Authority: Briefing Notes.' *Sarawak Gazette* 106(1463): 12–15.

Ngidang, D., 1997. 'Native Customary Land Rights, Public Policy, Land Reforms and Plantation Development in Sarawak.' *Borneo Review* 8(1): 63–80.

————, 1998. 'People, Land and Development: Iban Culture at a Crossroad.' Paper presented at the 'Second National Smart Partnership Dialogue', Kuching.

————, 2000. 'A Clash Between Culture and Market Forces: Problems and Prospects for Native Customary Rights Land Development in Sarawak.' In M.B. Leigh (ed.), op. cit.

————, 2001. 'Key Issues and Challenges in Native Customary Land Development: The Sarawak Experiences.' Paper presented at the 'Sabah Native Land Conference', 30 June–1 July.

Porter, A.F., 1967. *Land Administration in Sarawak: An Account of Development of Land Administration in Sarawak from the Rule of the Raja Brooke to the Present-Time (1841–1967)* . Kuching: Government Printer.

Richards, A.J.N., 1961. *Land Law and* Adat *in Sarawak* . Kuching: Government Printer.

Sarawak Government, 1959. *Land and Survey Department Annual Report*. Kuching: Government Printer.

————, 1997. *Handbook on The New Concept of Development of Native Customary Rights (NCR) Land*. Kuching: Government Printer.

Sather, C., 1990. 'Trees and Tree Tenure in Paku Iban Society: The Management of Secondary Forest Resources in a Long-Established Iban Community.' *Borneo Review* 1(1): 16–40.

Songan, P. and A. Sindang, 2000. 'Identifying the Problems in the Implementation of the New Concept of Native Customary Rights Land Development Project in Sarawak Through Action Research.' In M.B. Leigh (ed.), op. cit.

Zainie, Z.K., 1994. 'Native Customary Land: Policies and Legislation.' Paper presented at a seminar on 'Native Customary Land', Kuching, 29 September–3 October.

Chapter Four

Decentralisation, Forests and Estate Crops in Kutai Barat District, East Kalimantan[1]

Anne Casson

Introduction

This chapter examines the initial impacts of decentralisation on forests and estate crops in the district of Kutai Barat, East Kalimantan. It is one of nine district-level case studies carried out during 2000 and early 2001 by the Centre for International Forestry Research (CIFOR) in four provinces: Riau, East Kalimantan, Central Kalimantan, and West Kalimantan. Fieldwork for this study was conducted in mid-2000 and the author has relied on secondary material and key informants to update some information.

Located in the upper region of East Kalimantan's Mahakam river basin, West Kutai district was formed through the administrative division of the previous Kutai district shortly after the Habibie government issued Law No. 22/1999 and Law No. 25/1999 on the decentralisation of authority from the central government to district governments. As a newly formed district, Kutai Barat had limited infrastructure and revenue. Local government officials also had limited capacity to develop policy and sustainably manage natural resources.

Decentralisation did, nevertheless, provide opportunities for the government of Kutai Barat to secure a greater portion of the revenues generated by forests and mineral resources extracted within the district; and to build up the district's physical infrastructure and industrial facilities. In 2000, the Kutai Barat district government issued large numbers of small-scale timber extraction licenses,

[1] Research for this chapter was supported by the Centre for International Forestry Research, the Australian Centre for International Agricultural Research and the United Kingdom's Department for International Development. The opinions expressed here are the views of the author and do not necessarily represent the official policies of any of these organisations. The research itself was supported and informed by the following people in East Kalimantan, Jakarta and Bogor: Anja Hoffmann, Ben Santoso, Eric Wakker, Frank Flashe, Gottfried von Gemmingen, Grahame Usher, Hans Beukeboom, Hery Romadan, Kadok, Liz Chidley, Neil Scotland, Rudi Ranaq, Rona Dennis, the staff of BIOMA, and all of the local inhabitants of Kutai Barat who gave their time while I was in the area. During the time of writing, additional support and encouragement was provided by Carol Colfer, Chris Ballard, Colin Filer, Ketut Deddy, Hidayat Al-Hamid, Peter Kanowski, Stephen Midgley, Yvonne Byron, Chris Barr, Daju Pradnja Resosudarmo, John McCarthy, Lesley Potter and Simon Badcock. Finally, I would like to thank Erna Rositah (BIOMA), who accompanied me in the field for three weeks. With her help and expertise, I learned a great deal and had many an adventure.

known as Hak Pemungutan Hasil Hutan (HPHH) permits, to establish a district regulatory regime for forest exploitation. The district government also indicated that it would encourage investors in the oil palm industry to establish operations in Kutai Barat. The HPHH scheme resulted in widespread and accelerated deforestation in the district and attracted considerable criticism from a wide range of stakeholders, including NGOs, donors, provincial governments and the forest industry. The district government eventually heeded these concerns and set about establishing a more equitable and sustainable forest regime, which seeks to facilitate local development.

Kutai Barat and Its Resources

Geography

Kutai Barat is one of the newly formed districts (*kabupaten*) in East Kalimantan. It was officially established in November 1999, in accordance with Law No. 47/1999, which outlined the division of the original *kabupaten* of Kutai[2] into three districts: Kutai Barat, Kutai Timur and Kutai Kartanegara[3] (Figure 4.1). Before Kutai was divided into three smaller *kabupaten,* it was the largest district in East Kalimantan, covering 94 629 square kilometres (km^2), or approximately 46 per cent of the province's total land area (BAPPEDA and BPS 1998). Kutai also had a long history as an administrative unit, having originated from the Kutai sultanate established late in the 15th century along the Mahakam River (Magenda 1991).

The decision to divide Kutai into three districts was long awaited, as the sheer size of the original district made it difficult to administer. Indeed, several remote areas, which now primarily lie within Kutai Barat, have limited physical infrastructure and industrial facilities due in part to their isolation from the former district's administrative centre. Moreover, the division followed the release of Indonesia's regional autonomy laws (No. 22/1999 and No. 25/1999), which ostensibly aimed to provide an opportunity for further autonomy in the region and to allow local governments to be more responsive to local communities (Bupati Kutai 2000).

[2] Kutai was established as a *kabupaten* in 1959 with the enactment of Law No. 27/1959 concerning the formation of Daerah Tingkat II in East Kalimantan. The first Regent of Kutai was A.R. Padmo.
[3] When I visited Kutai Barat in July 2000, Kutai Kartanegara was referred to as Kutai Induk, meaning the 'mother' district. However, the people and government of this district now prefer to call it Kutai Kartanegara. This term is therefore used throughout this chapter.

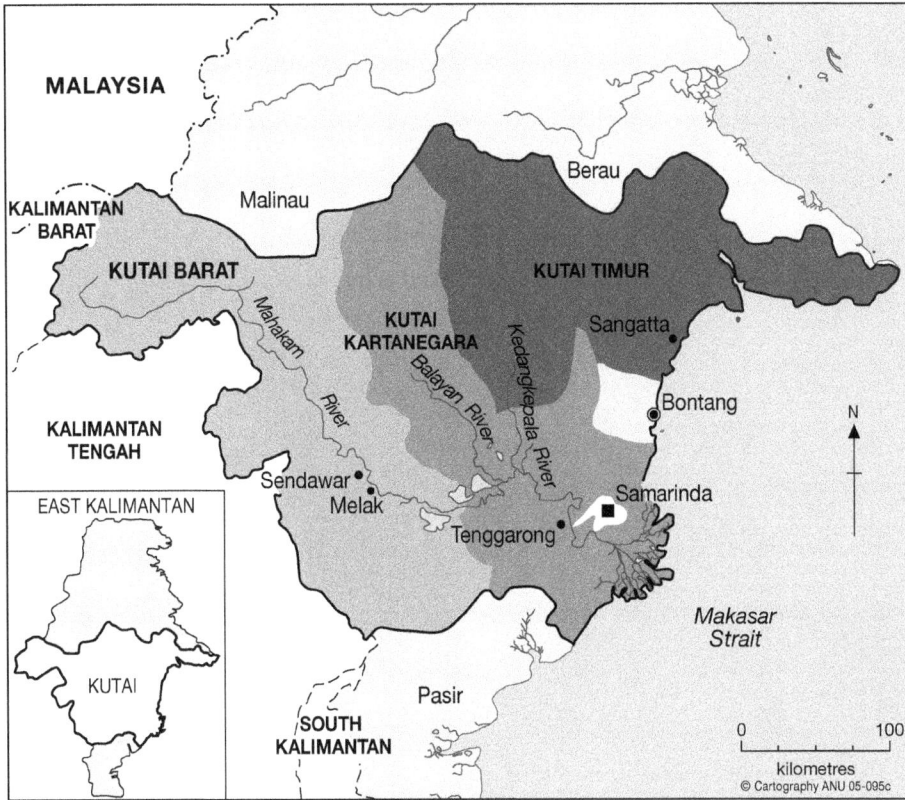

Figure 4.1. Kutai Region, East Kalimantan

Source: Map data from GTZ Sustainable Forest Management Project

Kutai Barat now spans an area of approximately 32 000 km², or 16 per cent of East Kalimantan's total land area. It is located in the western part of the province and borders both Central and West Kalimantan, as well as the East Malaysian state of Sarawak. The newly formed district consisted of 14 sub-districts (*kecamatan*), 205 villages and approximately 150 000 people in 2000[4] (Figure 4.2). The capital of Kutai Barat is Sendawar, but all of the existing government offices were located within the town of Melak. When fieldwork was undertaken in July 2000, Kutai Barat had a temporary Bupati, or district head, named Bp. Rama Alexander Asia. He had the difficult job of forming a new district government and legislative assembly without having full authority and legitimacy to rule. He was formally elected in March 2000 and the district legislative assembly was elected in December 2000.

---

[4] Statistics vary on the actual area and population of Kutai Barat. According to the Regional Development Planning Agency (BAPPEDA and BPS 1998), the area of Kutai Barat is 33 118 km² and the population is 122 153. However, Bupati Kutai (2000) claims that the area is 31 628 km² and the population is 150 871.

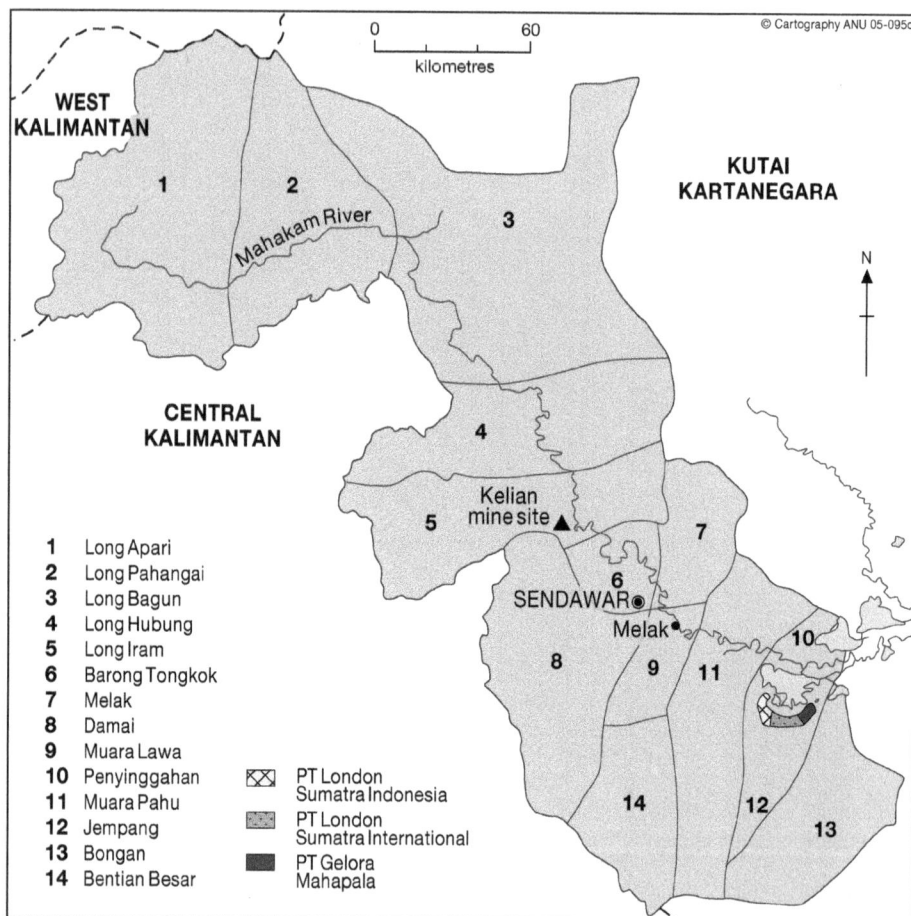

Figure 4.2. Kutai Barat District, East Kalimatan

Source: Map data from GTZ Sustainable Forest Management Project

Economy

Most of the physical infrastructure and industrial facilities established in the original district of Kutai are now located within the jurisdiction of a newly formed district now known as Kutai Kartanegara. This meant that Kutai Barat had fairly limited infrastructure, but perhaps no more so than some of the other newly formed districts in East Kalimantan, such as Malinau or Nunukan. It was also quite isolated. To get to Kutai Barat one had to catch a boat from Samarinda up the Mahakam River. Depending on the boat, the trip took anywhere between eight and 24 hours. There was one asphalt road in Melak that was 25 km long. The rest of the roads were dirt roads and almost impassable during the wet season. In fact, five sub-districts (Long Apari, Long Pahangai, Long Bagun, Long Hubung and Penyinggahan) could not be reached by road in 2000.

Government offices in Kutai Barat also had limited facilities and were poorly resourced.[5] In fact, many offices had not been established when fieldwork for this study was conducted.[6] By June 2000, for instance, there was no district government forestry agency, only a branch office of East Kalimantan's Provincial Forestry Service (Cabang Dinas Kehutanan). Public servants working in the district also openly admitted that they were poorly trained and lacked the knowledge required to run a district and develop regional policy (interviews with Kutai Barat government officials, 28 July 2000). The closest university was in Samarinda, and there were just two secondary schools — one in Melak and the other in Long Iram. Other community services in the district were also extremely poor. For instance, there was no hospital and no reliable telephone or electricity supply.

Forest Resources

Kutai was once covered in dense tropical forest. These forests were described by the Norwegian naturalist and explorer Carl Bock, who was commissioned by the Dutch colonial government in the 1880s to travel halfway up the Mahakam River. In a report on the journey, Bock wrote:

> Enormous trees, with massive straight stems rising sixty or eighty feet from the ground before throwing out a single branch, overshadowed the rank vegetation beneath, the thickness of which rendered it impossible to penetrate into the forest more than a few yards from the riverside (Bock 1985: 51).

Since the Suharto government opened up Indonesia's outer island forests to large-scale cutting in the late 1960s, most of the forest described by Bock has been cleared (Bupati Kutai 2000). Potter (1990) estimates that around forty per cent of Indonesia's log production originated from East Kalimantan during the period 1970–79. A large proportion of this timber came from the area now known as Kutai Kartanegara, just west of Kutai Barat, because the largest stands of commercial species, such as *meranti*, *keruing* and *agathis*, could be found there. The Mahakam River also provided a well-developed transport system (Manning 1971).

In addition to the impact of large-scale logging, extensive areas of forest land have been converted to plantations or agriculture (Bupati Kutai 2000). A large proportion of Kutai Kartanegara's forest cover was also severely burnt during

---

[5] It is perhaps interesting to note that most of the office furniture in the Regent's office had been donated by PT Kelian Equatorial Mining, a company mining gold in the area.

[6] When I visited in July 2000, the only offices to be found in Melak were: *Cabang Dinas Kehutanan* (branch office of the Provincial Forestry Service), *Dinas Pertanian* (District Agriculture Office), *BAPPEDA* (Regional Planning Agency), *Dinas Pendapatan Daerah* (District Income Office), *Sospol* (District Social-Politics Office), *Dinas Kesehatan* (District Health Office) and *Dinas Perkebunan* (District Estate Crops Office).

the 1982–83 and 1997–98 forest fires (Brookfield et al. 1995; Hoffmann et al. 1999). In 2000, the landscape consequently bore little resemblance to that described by Bock. Few trees could be seen on a journey up the Mahakam River and large sawmills dominated the landscape between Samarinda and Tenggarong.

Before Kutai was divided into three districts it had 1.8 million hectares (ha) of forest land classified as 'Protected Forest'; 270 000 ha of 'Parks and Reserve Forest'; 2.6 million ha of 'Limited Production Forest'; 3.3 million ha of 'Production Forest'; 3.1 million ha of 'Conversion Forest'; and 22 724 ha of 'Research Forest' (BAPPEDA and BPS 1998). While these categories do not necessarily correspond with actual forested area, the *Forest Land Use Consensus Plan* (*Tata Guna Hutan Kesepakatan*) was developed to show definitive boundaries between the various categories of land under the Forestry Department's control. The plan was revised in the mid-1990s to reconcile provincial and district needs, including a desire to convert production forest to conversion forest in order to facilitate oil palm developments. This revision resulted in the development of *Provincial Land Use Plan* (*Rencana Tata Ruang Wilayah Propinsi*) classifications in the mid-1990s. A *District Land Use Plan* (*Rencana Tata Ruang Wilayah Kabupaten*) was also drawn up for the original administrative boundary of Kutai, but had not been completed for Kutai Barat when fieldwork for this study was undertaken. However, the district government estimated that approximately 50–60 per cent of Kutai Barat was still forested in 2000 (Bupati Kutai 2000).

Before Kutai was divided into three districts, the Mahakam Ulu branch office of the Provincial Forestry Service monitored forest activities and production in the area now known as Kutai Barat. The forest area monitored by this agency has remained more or less the same since the partition of Kutai. According to its own statistics, the Mahakam Ulu area produced approximately 3 million cubic meters (m³) of logs during the period 1994–98. This made the area the fourth largest producer of logs within the province after Mahakam Tengah (now known as Kutai Kartanegara — 7.3 million m³), Berau (3.9 million m³), and Bulungan Utara (3.6 million m³). According to official statistics, log production in the Mahakam Ulu area had gradually declined between 1995 and 2000 from 818 324 m³ in 1994/95 to 619 426 m³ in 1998/99 (Kalimantan Timur 1999). However, many suspected that there had been an increase in illegal logging in the area during the same time period (interviews with various NGOs based in Samarinda and staff at the district forestry office in Melak, Kutai Barat, July 2000). Official statistics were therefore likely to understate real timber production from the area. Growing volumes of timber were expected to come out of the Kutai Barat area in the near future because much of Kutai Kartanegara has already been logged out and badly affected by the 1997–98 forest fires (Hoffmann et al. 1999).

In 1998/99, 22 companies had been granted timber concessions known as Hak Pengusahaan Hutan (HPH) within the Mahakam Ulu region, covering a total area of approximately 2.6 million ha. However, by the time that Kutai Barat was established as a *kabupaten*, there were only 10 active HPH concessions in the region, covering a total area of approximately 1.6 million ha. Most of these companies were operating in the far reaches of the district where there was very little physical infrastructure and few people. Two of the 10 active HPH companies — PT Kemakmuran Berkah Timber and PT Daya Besar Agung — were working together with the state-owned forestry companies, Inhutani I and Inhutani II respectively. In 1998, these 10 HPH-holders produced approximately 346 000 m³ of logs. This was roughly half of Mahakam Ulu's total annual log production for the year 1997/98 and approximately 15 per cent of Kutai's annual log production.

In addition to these 10 companies, five HPH companies had requested extensions of their concession licenses and were expected to become active within the next few years (BAPPEDA 1997; Kalimantan Timur 1998b). These five companies were expected to operate over a total area of 442 500 ha. Four of these companies were working together with Inhutani I. It is possible that much of this area may also be logged before other districts with high timber potential, such as Berau, because timber originating from the Kutai Barat region can more easily be transported to one of the many mills that line the Mahakam River between Tenggarong and Samarinda.

Seven companies had also been granted concession licenses in Kutai Barat to develop industrial timber estates, or Hutan Tanaman Industri (HTI). Areas allocated to these seven plantation companies covered a total area of 119 827 ha (Kalimantan Timur 1998a). Most of these plantations were located in the southeastern region of Kutai Barat. Two of these HTI companies — PT Riau Timas and PT Marimun Timber — were private timber estate companies, which operated independently. The remainder were participants in the government's HTI-Trans program, initiated by the Suharto government to provide employment for transmigrants from Java and other more populated parts of Indonesia. These estates were established on areas formerly managed as HPH timber concessions. Most of the HTI plantation companies were run by timber companies operating in the area; however, PT Alas Cakrawala was managed by Inhutani I. In 2000, only 23 914 ha had been planted to timber estates in Kutai Barat, amounting to 20 per cent of the total concession area allocated. Most of the planted area fell within concessions managed by three companies: PT Anangga Pundinusa, PT Hutan Mahligai and PT Kelawit Wana Lestari.

Agro-Industrial Estate Crops

As of 1999, the three main agro-industrial plantation crops cultivated within the original administrative boundary of Kutai were oil palm (40 164 ha), rubber (33 935 ha) and coconut (20 109 ha) (Kalimantan Timur 1998a). Rubber and coconut are traditional crops and most of the estates were owned and managed by smallholders in 1998. In contrast, oil palm estates were a relatively new development, and all of the estates had been established by private companies since 1990 (ibid.). Within the original administrative boundary of Kutai, most oil palm development had occurred in Kutai Timur, followed by Kutai Barat and Kutai Kartanegara. All three regions had ambitious plans to develop the sector and many companies had already received location permits, especially in the districts of Kutai Timur and Kutai Barat. A considerable area of forest land designated for conversion (more than 600 000 ha) had also been released for oil palm development in the district of Kutai Timur. However, little forest land had been released in Kutai Barat or Kutai Kartanegara (Figure 4.3).

In Kutai Barat, there were only three estates in which oil palm had already been planted by 2000: PT London Sumatra International, PT London Sumatra Indonesia and PT Gelora Mahapala. All three of these estates fell under a company known as PT London Sumatra International Tbk (PT LonSum).[7] PT LonSum had a significant presence in the district and was surrounded by conflict and controversy. The company had established plantations on land belonging to a number of Dayak Benuaq villages located in the Lake Jempang area (Figure 4.3), and it had been heavily criticised by grassroots organisations and NGOs for its alleged association with the 1997–98 forest fires and illegal land clearing; and for its oppressive action against local people (Muliastra et al. 1998; Gönner 1999; Ruwindrijarto et al. 2000; Wakker et al. 2000). Carrying substantial debts since the Asian financial crisis struck in mid-1997, the company's finances have also attracted a lot of interest both nationally and internationally (EIA 1998; Wakker 1999; Wakker et al. 2000).

---

[7] PT London Sumatra was founded in 1906 by a British company, but later became a subsidiary of British palm oil traders Harrisons and Crossfield. Harrisons and Crossfield was established in 1844 as a wholesale dealer in coffee and tea. The company later became active as managing agents and plantation proprietors in Ceylon and Malaysia, and unusually high profit from rubber encouraged further expansion to include Sumatra's East Coast in 1907. Several concessions were then leased in Deli for tea, coffee, rubber and tobacco, and in 1909 operations were extended to Java, where two more subsidiaries were formed. While the exact figures concerning Harrisons and Crossfield's holdings during these early years in the 1920s are unavailable, the company's landholdings were estimated to be around two million acres in Malaya, North Borneo, Indonesia, India, Ceylon and East Africa (Stoler 1985). In the mid-1990s, Harrisons and Crossfield sold its stake in the company and LonSum became an Indonesian company publicly listed on the Jakarta and Surabaya stock exchanges. Shareholders in the company included a number of prominent Indonesians with close connections to the Suharto family, such as Ibrahim Risjid, Andry Pribadi, and Henry Liem. Ibrahim Rashid was also one of the founders of the Salim Group — one of Indonesia's largest Chinese-Indonesian owned companies that has close ties to the former Suharto government. He was also the founder and majority shareholder of the Risjadson Group.

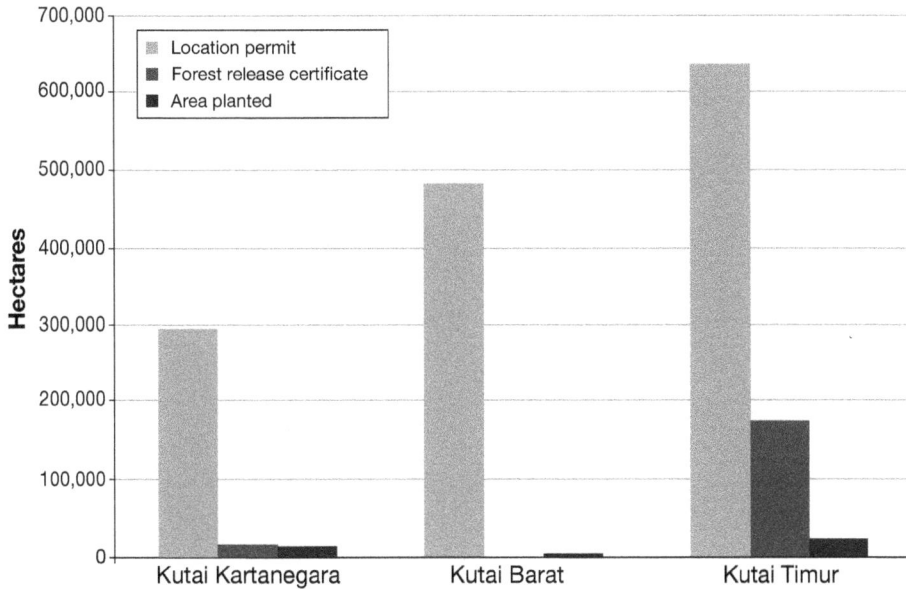

Figure 4.3. Actual and planned oil palm development in Kutai Kartanegara, Kutai Barat and Kutai Timur, March 2000

Source: Kalimantan Timur 1998a; unpublished statistics from the Kutai Kartanegara plantation office

At the beginning of 1998, LonSum management reacted to continuing adverse business conditions by deferring both the construction of a new oil palm mill in South Sumatra and planting activities in South Sumatra and East Kalimantan. In 1998, the company had only planted 99 ha on its new estates because the crisis had reduced the company's internal liquidity. When fieldwork was conducted in the area, the company had stopped all planting operations due to a lack of funding. Community conflict in the area had eased as a consequence, but other elements of the community were suffering. Not only had some local people lost their land to oil palm, but they had also lost any employment opportunities previously being offered by the company. Many had also joined the company's PLASMA scheme in which they had been promised two hectares of oil palm estate. By mid-2000, community members were just starting to realise that there was little chance of this scheme going ahead. Moreover, many of the oil palm trees had already started to bear fruit, but the company had no funds to establish a factory. The fruit was rotting on the ground or being fed to chickens as the nearest factory was in Pasir district, too far away to process oil palm fruit harvested from the LonSum sites.[8]

---

[8] Palm oil needs to be extracted from oil palm fruit within 24 hours.

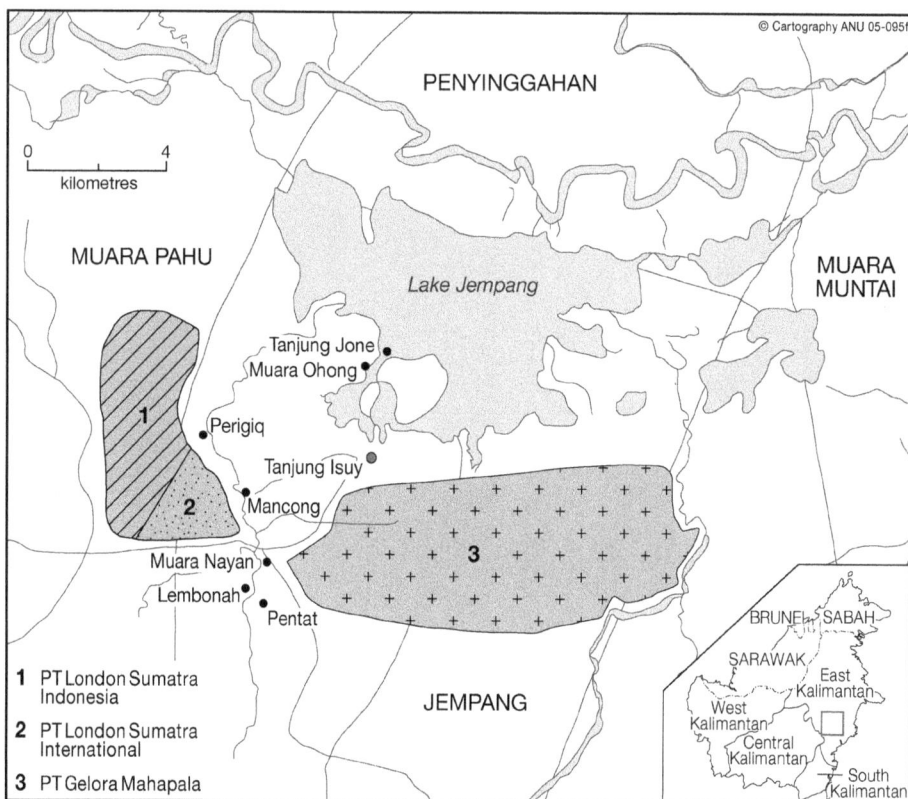

Figure 4.4. Location of villages in the PT London Sumatra Plantation Area, East Kalimantan

Source: Map data from GTZ Sustainable Forest Management Project[9]

Decentralisation in Kutai Barat

District Finances

Kutai Barat faces many challenges in the years ahead because it is the least developed region originating from Kutai district. Before the original district of Kutai was divided into three districts, it was the richest region in the province of East Kalimantan. Most of this income came from the district's natural resource base. In fact, Kutai has a long history of natural resource use and extraction. The modern founder of the Kutai sultanate (Sultan Mohammad Sulaiman, 1845–99) had a great talent in commercial activities, leasing out Kutai's lands for coal exploitation and plantations (Magenda 1991). Revenues also came from

[9] Official map data on the location of the PT LonSum plantations in the Lake Jempang area do not yet exist because the company has not acquired a number of permits from the Ministry of Forestry and the National Land Agency. The locations of the plantations shown on this map are therefore estimated from a visit to the area.

taxes levied on forest products transported down the Makaham River and royalties received from coal and plantation activities. Shortly after coal and oil were discovered in the sultanate in the late 1890s, a new sultan (Sultan Alimuddin, 1902–20) had to sign a treaty with the Dutch relinquishing certain rights over taxes. The treaty also stipulated that the Ulu Mahakam area, now known as Kutai Barat, would be governed by the Dutch. With the acquisition of the Ulu Mahakam area, the Dutch were able to control traffic along the Mahakam River and restrict the activities of the Kutai sultanate. With the imposition of these restrictions, the sultan received an annual salary of 25 000 guilders in addition to 50 per cent of oil royalties and 10 per cent of forest royalties (Magenda 1991).

During the timber boom of the late 1960s and 1970s,[10] the Kutai government again benefited from the exploitation of forest resources in the Mahakam region (Manning 1971; Magenda 1991). While the central government profited the most from extensive logging via taxes obtained from the HPH licence fee, a 10 per cent export tax, and a forest product royalty, the Kutai government still received a great deal from their own state-run timber companies (*Perusahaan Daerah*); as well as from taxes from timber and log pond retribution (Magenda 1991).

Before Kutai was divided into three districts, the district had a regional income of approximately Rp184 billion (or approximately US$18.4 million)[11] in 1997/98. This was more than double the income of other districts in the region. For instance, Pasir only had a regional income of approximately Rp68.7 billion, while Berau had a regional income of Rp55.2 billion (Kalimantan Timur 1998a). Since oil and gas were found in Kutai, the district's economy has been largely dominated by the oil and natural gas industry[12] (42.24 per cent), followed by the forestry and agriculture sector (25.24 per cent), the processing industry sector (13.72 per cent), the building sector (8.55 per cent), and the trade, hotel and restaurant sector (8.25 per cent) (Bupati Kutai 2000).

In light of the above, a considerable percentage of district revenue was generated from taxes obtained from the mining and forestry sectors in the years 1997/98 (Kalimantan Timur 1998a). For instance, Kutai received approximately Rp83.9 billion from the mining sector and Rp26.7 billion from the forestry sector. Again this was more than other districts received over the same period of time: Pasir only received approximately Rp4.6 billion from the forestry sector and Rp12.9 billion from the mining sector in the year 1997/98. Pasir did, however, generate more revenue (Rp1.4 billion) from the plantation sector than Kutai

[10] During the 1970s, East Kalimantan produced 30–40 per cent of the nation's timber exports and around two thirds of the total national timber supply (Magenda 1991).
[11] Although the Rupiah has fluctuated considerably against the US dollar since mid-1997, an exchange rate of Rp10 000 to the dollar is assumed throughout this chapter.
[12] Most of the oil and gas deposits lie within the area now designated as Kutai Kartanegara and Kutai Timur.

because the majority of the East Kalimantan's plantation estates fell within this district (ibid.).

After Kutai was divided, the provincial government expected Kutai Barat to generate the least revenue in the area because it was the least developed and most isolated region in the Kutai area. A paper released by the Bupati of Kutai Kartanegara shortly after Kutai was divided into three districts projected that Kutai Barat would generate just US$2.1 million in revenue in 2000, while Kutai Kartanegara and Kutai Timur were expected to generate US$5.6 million and US$6.2 million, respectively (Bupati Kutai 2000). This clearly highlighted the financial problems Kutai Barat was expected to experience in the new era of regional autonomy. The district government was also concerned that regionally generated revenues would not be sufficient to establish a new district, capital city and the entire associated infrastructure (personal communication, head of the Regional Development Planning Agency, Kutai Barat branch, July 2000).

Realising this, the governments of Kutai Kartanegara and Kutai Timur pledged to provide Kutai Barat with Rp3 billion in order to finance some of the infrastructure needed to establish a new district. The funds were to be used to establish the necessary government offices as well as the houses for the Bupati, Assistant Bupati, the District Secretary, and heads of district government agencies (*Kaltim Post*, 16 November 1999).

The local government of Kutai Barat also began to make plans to be fiscally independent. The Bupati was keen to attract new investment to the region and to generate regional income from the district's natural resource base. He was also supportive of existing industries such as PT Kelian Equatorial Mining (PT KEM)[13] — a large open-cut gold mine located in the sub-district of Long Iram, which had long attracted criticism from national and international NGOs.[14]

PT KEM is one of the largest gold plants constructed in the world and it had the potential to generate substantial revenue for the district. Between 1995 and 2000, the company had paid a total of Rp85 billion to the central government and Rp144 billion to the East Kalimantan provincial government.[15] Most of this revenue came from corporate income tax and land rent tax. Regional autonomy gave the Kutai Barat government the opportunity to receive a large proportion

---

[13] The company is an Indonesian incorporated company, 90 per cent owned by Rio Tinto Ltd and 10 per cent by PT Harita Jayaraya of Indonesia. The Kelian deposit was discovered in 1976 by Rio Tinto's exploration group, but commercial production did not begin until January 1992. Once operational, the mine was producing and exporting an average of 450 000 ounces of gold and 400 000 ounces of silver per annum.

[14] These NGOs have claimed that the mine has a substantial environmental and social impact in the area. Mine by-products are believed to be polluting local waterways and the surrounding environment; and NGOs have claimed that up to 80 000 people were forcibly removed from the site and resettled. This has given rise to human rights abuse issues and complicated compensation claims (Down to Earth 1998).

[15] Much of this revenue subsequently went to the central government.

of these taxes. In fact the Bupati had already had several discussions with the mine about how taxes would be distributed between the Kutai Barat government and the provincial and central governments (personal communication, PT KEM management, July 2000). Nevertheless, problems existed because NGOs and some local communities did not support the mine's activities. Revenue generated from the mine was also limited by the fact that the company planned to close down its operations in 2004. The planned closure was causing local government officials and communities much concern as the region only had the potential to benefit from tax payments for two to three years. The district government would then be left with the consequences of the company closing down, unemployment and environmental damage[16] caused by mining activities.

Creating a District Regulatory Regime to Exploit Forest Resources

While Kutai Barat had limited infrastructure and revenue it was rich in natural resources, particularly forest resources. Realising that revenue from the PT KEM mine was limited, the local government began to identify ways to facilitate and support local development through forest exploitation shortly after it was formed. According to the Kutai Barat district government, the central government had given district governments some rights to generate revenue from forest resources through the issuance of the new decentralisation laws (No. 22/1999 and No. 25/1999); and Regulation No. 6/99 on *Forest Utilisation and Forest Product Harvesting in Production Forests*. Following the release of the latter regulation, the Kutai Barat government followed the lead of the Kutai Kartanegara government to issue 100 ha concessions as a means to generate revenue and allow local people to benefit from forest resources. By the end of 2000, the Kutai Barat government had issued 220 Permits to Use and Harvest Timber (Ijin Pemanfaatan dan Pemungutan Kayu) and nearly 50 HPHHs (Forest Product Harvesting Rights). These licences cumulatively covered approximately 22 300 ha of forest land and allowed the Kutai Barat government to generate approximately US$27 000 in revenue (*Kaltim Post*, 24 August 2000). In 2000, the HPHH scheme had gained a lot of support from the general populace, primarily because it was realised that HPHHs could generate substantial revenue for the new district, and also because the allocation of HPHH permits would enable local people to secure a greater share of the benefits from forest resources.

---

[16] PT KEM had admitted that it was unable to rehabilitate approximately 950 ha out of the total 1285 ha of forest land disturbed by mining operations. This land would instead be turned into lakes and wetlands after mine closure. To compensate for the land it is unable to rehabilitate, PT KEM is currently planting trees in two replacement areas: Linggau Plateau, which is within the Contract of Work area, and Bukit Suharto, north of Balikpapan. Around 5700 ha of forest land remain intact in the concession area, and the PT KEM management group hoped that its status could be changed to 'Protected Forest' area after the mine closed (personal communication with PT KEM management, July 2000).

However, a number of problems were also noted. For instance, it was recognised that there was potential for local people to be exploited through the process, particularly as they were dependent on outside partners for capital and machinery. Local government also lacked the capacity to ensure that forest exploitation was carried out in a sustainable and equitable manner. The small-scale licences were undoubtedly fuelling an increase in deforestation and little thought had been given to forest rehabilitation or reforestation. Consequently, the Mahakam River was flooded with timber (*dibanjiri kayu*) from the region — one woman interviewed along the Mahakam River between Tenggarong and Samarinda said she had not seen so much timber pass through the region since the 1970s.

Donors and non-government organisations raised a number of concerns about the issuance of these new permits. They raised concerns about the potential for HPHHs to increase conflict among communities and individuals — especially since many of the permits were being allocated in HPH concessions, HPHH rights could be manipulated by government officials and outside interests, and their issuance could give negative connotations to the decentralisation process.

The timber industry also raised concerns about the fact that HPHHs were being issued within existing HPH concessions, and this was a matter of concern to provincial and central governments as both were under significant pressure from the timber industry to restore law and order in the forest estate. For instance, in early 2000, 77 loggers in Kutai and Bulungan threatened to close down their operations if local and provincial governments did not prevent local people from logging their concessions and disrupting their activities (*Jakarta Post*, 21 February 2000). The Association of Indonesian Timber Concession Holders (Asosiasi Pengusaha Hutan Indonesia) also complained that several companies in Kutai and Bulungan were unable to continue operating as local peoples had seized their heavy equipment to demand payments amounting to billions of rupiah (*Jakarta Post*, 21 February 2000: 9). Moreover, in February 2000, some 10 foreign investors and plywood buyers, mainly from South Korea, threatened to pull out of their contracts due to concern over escalating conflicts between timber companies and local people. The buyers said that they were worried that plywood mills would not be able to meet delivery schedules as many timber companies had stopped logging operations as a result of prolonged disputes with local people (ibid.).

In early 2000, the central government realised that it had lost control over the allocation of HPHH permits and was losing large amounts of potential revenue from district timber regions. The Ministry of Forestry and Estate Crops then suspended its earlier regulation giving district governments the right to allocate small-scale logging permits. Both the Bupati of Kutai Kartanegara and Kutai Barat were ignoring this decision and arguing that, under regional autonomy, they were duty bound to ensure that local people would directly benefit from forest resources (personal communication, Bupati of Kutai Barat and Kutai Kartanegara, 5 August 2000).

A New Forestry Vision — Community Forestry

Heeding the concerns raised by the forest industry, donors and local non-government organisations, the newly elected Bupati, Rama Alexander Asia, eventually decided to slow down the issuance of HPHH licences in 2001. He also solicited assistance from organisations such as the USAID-funded Natural Resource Management Project, the GTZ Sustainable Forest Management Project, the International Centre for Research on Agroforestry, and local NGOs such as PLASMA and Puti Jaji, to develop a more sustainable model of community-based forest management in the area.

In November 2000, a West Kutai Regional Forestry Program Working Group (Kelompok Kerja Program Kehutanan Daerah or KKPKD) was formed with assistance from the Natural Resource Management Project. Representing a diverse set of local forest sector stakeholders including government officials, *adat* or customary rights leaders, community representatives, NGOs, university professors and the private sector, KKPKD sought to understand the complexities of forest management in West Kutai District and to develop transparent and accountable management strategies for balancing sustainable forest management and improved community welfare in a participatory manner. In recognition of the importance of KKPKD's work in terms of good governance and sustainable forest management, the Bupati of West Kutai formalised this working group on 2 January 2001 with Decree No. 453/K.065/2001. This provided a clear mandate for KKPKD 'to establish a viable forestry program involving forestry-related activities and stakeholders in West Kutai District' (KKPKD 2001).

Over the course of 2001, KKPKD worked through a number of planning activities in a wide range of formal and informal workshops and meetings, both large and small, to first understand forest management issues and to then develop a realistic forest management plan. While these major activities were being carried out, KKPKD also sought to make the results transparent to the general public. Innovative public consultation processes and regular use of electronic media — particularly radio — kept the public informed and involved (KKPKD 2001).

The West Kutai District Forest Management Plan was divided into seven main themes: forest recovery and management, institutional development, law enforcement, infrastructure, regional policy development, recognition and empowerment of *adat* customary and community rights, and human resource development. Within these seven main themes, 52 specific activities were presented and ranked in terms of their relative priority and targeted for completion within the next ten years. The plan also included information on supporting institutions and their roles, time frames, resources, basic assumptions, expected results, and criteria for the success for each activity (KKPKD 2001).

In 2002, two important regional or district regulations (Peraturan Daerah or Perda) were developed in accordance with this plan — Regulation No. 18/2002 on forestry and a draft regulation on the implementation of community forestry in the district. The regulation on forestry, which was signed by the Bupati on 4 November, provided the umbrella framework for all forestry matters in the district, including community forestry. Article 9 of this regulation stated that the local government would recognise and gazette customary forest (*adat*) areas. It also stated that the local government would allow local people to manage forest resources in accordance with their customary rules and regulations. To strengthen this point, an entire chapter (Chapter 5) was devoted to the rights and roles of local people in local forestry matters. Forest planning was to be carried out in a transparent, participatory, accountable, integrated fashion, which allowed for community aspirations (Article 13); and local people would be kept well informed about large-scale forest exploitation (Article 18). The participation of local people in conservation and forest rehabilitation was also stressed in Articles 28 and 43.

The draft regulation on community forestry further strengthened the local government's intent to recognise and legitimate customary forest management systems. Article 2 of this regulation stipulated that the government would respect and recognise local communities that carried out forest management in an equitable and sustainable manner. If this is proven to be so, the government will recognise various types of community forestry, including customary forest management, village forest management, and local conservation efforts (Article 3). In order for the local government to grant a right to carry out community forestry, the designated area must be proven to be free from conflict and already covered by internal regulations which sustain forest resources (Article 7). If granted a community forestry licence, the community is obligated to sustainably manage forest resources in a just and equitable fashion (Article 23).

However, while both of these regulations promote more equitable forest exploitation, several articles contradict central government regulations and could be considered to be illegal. For instance, in the regulation on forestry, the Bupati retains the sole right to issue licences for timber collection, community forestry, forest use by the wood industry, non-timber forest product collection, and the use of environmental services (Article 34). This article contradicts two key central government regulations — Regulation No. 34/2002 on *Forest Planning, Management and Use* (the long awaited clarifying legislation of the *Basic Forestry Law* No. 41/1999) and Decree No. 31/2001 on *Community Forestry*. The regulation and the decree both stipulate that the Bupati can only issue these permits after he has sought approval from the Ministry of Forestry.

Similarly, in the draft regional regulation on community forestry, the Bupati retained the sole right to issue a 'Community Forestry Permit' (Izin Kehutanan Masyarakat) after he received a recommendation from the head of the District Forest Department (Article 6). This contradicted Decree No. 31/2001 which stipulates that the Bupati can only recommend that a 'Community Forestry Activity Permit' (Izin Kegiatan Hutan Kemasyarakatan) be allocated after he has sought permission from the Ministry of Forestry through the Provincial Governor. More significantly perhaps, the draft regulation specified that a Community Forestry Permit could be allocated for a period of 100 years (Article 19). However, Decree No. 31/2001 states that a Community Forestry Permit can only be allocated for 25 years. Given the above legislative contradictions, the Kutai Barat government needed to gain broad political support for district government regulations pertaining to forestry in order to avoid conflict with the central government, and to gain its support for their future forest management plans.

Decentralisation and the Oil Palm Sub-Sector

When fieldwork was undertaken in mid-2000, the local governments of Kutai Kartanegara and Kutai Barat were keen to develop the oil palm sub-sector as they saw it as a potential revenue generator. However, they were becoming increasingly frustrated with the central government, which implemented a moratorium on further forest conversion for plantation development in 1998 (Casson 2000). At a meeting held in Samarinda, the Assistant Head of East Kalimantan's Provincial Legislative Assembly publicly stated that provincial and district governments would no longer allow the central government to limit their ability to establish and promote further oil palm developments in the area:

> If the government continues to insist that they will not give out forest release permits to companies or communities in East Kalimantan, then the Kutai government is ready to seize them (translated from *Kaltim Post*, 1 May 2000).

He then went on to say that carrying out autonomy in the current era of globalisation is a great opportunity for East Kalimantan to move forward and open one million hectares of forest land for plantations, be it oil palm or other crops (ibid.).

In order to open one million hectares for plantation development, the Assistant Head of the East Kalimantan Legislative Assembly said that the provincial and district governments would seize control of forest release permits in order to facilitate development. He further said that they would legitimise this process by issuing their own legislation on the release of forest land for plantation development (*Kaltim Post*, 1 May 2000). The Bupati of Kutai Kartanegara was very supportive of these actions, stating that there were over 200 entrepreneurs waiting to invest in the sector. Many of these entrepreneurs were said to be from overseas countries, including Germany, Japan, Korea, Singapore and Malaysia (*Kaltim Post*, 28 July 2000).

To facilitate the development of oil palm plantations in the area, the National Land Agency (Badan Pertanahan Nasional) at both the provincial level and Kutai Kartanegara district level had discussed ways in which they could speed up the allocation of permits for oil palm estates. During an interview, the head of the provincial office of the National Land Agency explained that his office planned to lobby the provincial assembly to pass legislation that would enable them to issue location permits and land-use rights. Forest release permits would pass through the Dinas Kehutanan office at the district and provincial levels to the Governor of East Kalimantan. It was hoped that the Governor would be given the authority to release forest land in the province and pass on his recommendation to central government. By cutting the central government out of the permit-allocation process, the provincial and district governments expected to be able to accelerate oil palm development and forest conversion. At the time, the central government appeared to be willing to accommodate these plans, as a meeting was held in March 2001 between the National Land Agency and the National Development Planning Agency (Badan Perencanaan Pembangunan Nasional) to review the 1960 *Agrarian Law* in an attempt to simplify the permit-allocation process for potential investors (*Jakarta Post*, 2 March 2001). According to the provincial head of the National Land Agency potential investors were optimistic about these plans and eagerly awaited their implementation (personal communication, August 2000).

The Kutai Barat district government also wished to facilitate further oil palm development, but it was concerned about the fact that PT LonSum had failed to build a palm oil processing plant and had stopped further planting in the area. The district government was also aware that conflict surrounding PT LonSum's plantations was deterring other investors from developing oil palm plantations in the region. In order to attract investment, the district government was trying to secure the company's land use permit and to support the establishment of a palm oil processing factory (personal communication, Kutai Barat District Government Secretary, 28 July 2000). While it looked certain that the land use permit would soon be issued, it seemed unlikely that LonSum would be able to establish a factory or resume planting in the near future.

The Bupati of Kutai Barat was of the opinion that problems arising from the LonSum development were a consequence of central government control over the licensing procedures. He felt that the central government did not adequately consult or inform the local people about the company's development plans and that this had resulted in a number of misunderstandings which led to conflict and the occupation of the base camp. He was confident that decentralisation could provide district governments with the opportunity to better manage conflict with local communities because they would be more responsive to local needs and more aware of problems arising from such developments. He also felt that local government would better understand the concerns of local communities and be more able to accommodate these concerns into development plans (personal communication, Bupati Rama Asia, July 2000).

A number of international donor organisations and local NGOs held a similar opinion. Most thought that decentralisation had the potential to allow a lot more community consultation and participation in local government decision making. They were of the opinion that the central government had proved that it could not manage natural resources sustainably, and they were confident that decentralisation could improve the situation if implemented in a manner that supports good governance (Usher 2000). Many of the Samarinda and Balikpapan NGOs had committed themselves to helping the Kutai Barat government to build up their expertise, skills, knowledge and revenue (GTZ 2000). Some, such as Plasma, were also helping the district government to draw up district regulations. It remains to be seen if the local government of Kutai Barat has the skills and expertise to manage natural resources. There is always the danger that the Kutai Barat government will speed up the forest conversion process in order to facilitate further oil palm development and generate much-needed district revenue.

Conclusion

In Kutai Barat, decentralisation has the potential to build up the physical infrastructure and industrial facilities of the district. Before Kutai was divided into three regions, most of the original districts development occurred within the area now encompassed by the district of Kutai Kartanegara, while the outer regions of Kutai Barat were more or less neglected. Following the partition of Kutai, the Kutai Barat government has the opportunity to focus on developing these previously neglected areas and ensuring that more funds are directed to building up local infrastructure. A Kutai Barat government is also likely to pay more attention to local people and to listen to the needs and concerns of the local population.

However, the Kutai Barat government faces many challenges in the years ahead as it has to work with limited infrastructure, poorly skilled government officials and little revenue. PT KEM planned to close down its operations in 2004 and PT LonSum had limited funds to continue its operations and build a crude palm oil processing factory. Because of this situation, the local government was being forced to find new ways of generating income from its natural resource base, particularly forest resources. An example of this was the new HPHH scheme that enabled local government to generate income and local communities to benefit from the district's forest resource base. While this scheme may have been preferable to the former system whereby large HPH timber concessions were issued to conglomerates close to the Suharto family, the allocation of large numbers of HPHH permits resulted in increased social conflict and environmental damage. This was primarily because Kutai Barat's district government did not have the skills, staff or expertise to manage or monitor the scheme in 2000.

In 2001, the Bupati of Kutai Barat heeded concerns raised about HPHH licences by NGOs and donors. The Kutai Barat government slowed down the issuance of these permits and decided to develop more transparent and accountable management strategies for balancing sustainable forest management with improved community welfare. These strategies sought to facilitate and develop community forestry mechanisms in order to promote more equitable forest resource exploitation. Nevertheless, problems did exist because district regulations formalising these new arrangements contradicted various central government forest regulations. Agreements between both levels of government were therefore needed to ensure that local government regulations promoting more equitable sharing of forest resources could be implemented and accepted by multiple stakeholders.

References

BAPPEDA (Badan Perencanaan Pembangunanan Daerah). 1997. 'Data Pokok Pembangunan Daerah Kabupaten Daerah Tingkat II Kutai Tahun 1996/1997 [Basic Data on the Development of the Level II District, Kutai, for the Year 1996/1997].'

Bock, C., 1985 (1881). *The Head-Hunters of Borneo: A Narrative of Travel up the Mahakam and down the Barito, also Journeys in Sumatra.* Singapore: Oxford University Press.

Brookfield, H., L. Potter and Y. Byron, 1995. *In Place of the Forest: Environmental and Socio-economic Transformation in Borneo and the Eastern Malay Peninsula.* Tokyo: United Nations University Press.

Bupati Kutai, 2000. 'Paparan Bupati KDH Tingkat II Kutai Tentang Pemekaran Wilayah Kabupaten dati II Kutai Menjadi 3 (Tiga) Kabupaten dan 1 (Satu) Kota [The Explanation of the District Head of Kutai Concerning the Subdivision of Kutai to become Three Districts and One Town].' Samarinda: Bupati Kutai.

Casson, A., 2000. 'The Hesitant Boom: Indonesia's Oil Palm Sub-sector in an Era of Economic Crisis and Political Change.' Bogor: Centre for International Forestry Research (Occasional Paper 29).

Down to Earth, 1998. 'PT KEM Community Demands.' Viewed 1 March 2006 at http://www.gn.apc.org/dte/Cklcd.htm

EIA (Environmental Investigation Agency), 1998. *The Politics of Extinction: The Orang-Utan Crisis: The Destruction of Indonesia's Forests*. London: EIA.

Gönner, C., 1999. 'Causes and Effects of Forest Fires: A Case Study from a Sub-district in East Kalimantan, Indonesia.' Draft paper for the ICRAF methodology workshop on 'Environmental Services and Land Use Change: Bridging the Gap between Policy and Research in Southeast Asia', Chiang Mai, 31 May–2 June.

GTZ (Gesellschaft für Technische Zusammenarbeit), 2000. 'Paparan Ketua BAPPEDA Kabupaten Kutai Barat di Rio Tinto Bingung Baru [The explanation of the Head of BAPPEDA, West Kutai district regarding new confusions at Rio Tinto].' Paper presented to a planning workshop, Samarinda, 24 March.

Hoffmann, A., A. Hinrichs and F. Siegert, 1999. 'Fire Damage in East Kalimantan in 1997/98 Related to Land Use and Vegetation Classes: Satellite Radar Inventory Results and Proposal for Further Actions.' Samarinda: Ministry of Forestry and Estate Crops, GTZ and KfW (IFFM–SFMP Report 1a).

Kalimantan Timur, 1998a. 'Kalimantan Timur Dalam Angka [East Kalimantan in Figures].' Samarinda: Kantor Satistik [Provincial Statistical Office].

———, 1998b. 'Pemutahiran Data: Pengusahaan Hutan Tahun 1997/1998 [Corrected Data: Forest Companies for the Year 1997/98].' Samarinda: Kantor Wilayah Propinsi Kaltim [East Kalimantan Regional Office].

———, 1999. 'Dinas Kehutanan Dalam Angka Tahun 1998/99 [The Department of Forestry in Figures, 1998/99].' Samarinda: Propinsi Daerah Tingkat i Kalimantan Timur, Dinas Kehutanan [Government of the Province of East Kalimantan, Department of Forestry].

KKPKD (Kelompok Kerja Program Kehutanan Daerah), 2001. 'Program Kehutanan Kabupaten Kutai Barat [Forestry Program for the District of West Kutai].' Melak: KKPKD.

Magenda, B., 1991. *East Kalimantan: The Decline of Commercial Aristocracy*. Ithaca (NY): Cornell University.

Manning, C., 1971. 'The Timber Boom with Special Reference to East Kalimantan.' *Bulletin of Indonesian Economic Studies* 7(3): 30–60.

Muliastra, K., F. Rahmadani, S. Kasyanto, H. Suryani, 1998. 'Laporan Survey Kebakaran Hutan di Kalimantan Timur [Survey Report on the Forest Fires in East Kalimantan].' Jakarta: Report prepared for WWF Indonesia Program.

Potter, L., 1990. 'Forest Classification, Policy and Land-Use Planning in Kalimantan.' *Borneo Review* 1(1): 91–128.

Ruwindrijarto, A., C. Kirana, H. Pandito, H.R. Effendi, M. Minangsari, F. Ganswira, R.R. Sigit and R. Ranaq, 2000. 'Planting Disaster: Biodiversity, Social Economy and Human Right Issues in Large-scale Oil Palm Plantations in Indonesia.' Jakarta: Telapak, Madanika, Puti Jaji.

Stoler, A., 1985. *Capitalism and Confrontation in Sumatra's Plantation Belt, 1970–1979.* Michigan: University of Michigan Press.

Usher, G., 2000. 'Why Donors Should Concentrate their Programs at the Kabupaten Level, or Why Focusing Donor Support at the Kabupaten Level is More Likely to Achieve Better Natural Resource Management.' Samarinda: East Kalimantan Natural Resource Management Project.

Wakker, E., 1999. *Forest Fires and the Expansion of Indonesia's Oil Palm Plantations.* Jakarta: WWF Indonesia.

————, Telapak Sawit Research Team and J.W. van Gelder, 2000. 'Funding Forest Destruction: The Involvement of Dutch Banks in the Financing of Oil Palm Plantations in Indonesia.' Amsterdam: Report for Greenpeace Netherlands.

Part III. Local Interventions

Chapter Five

Community Mapping, Tenurial Rights and Conflict Resolution in Kalimantan[1]

Ketut Deddy

Introduction

Conflicts over land and natural resources often occur where there are overlapping resource interests among groups, communities or states. These overlapping interests usually become clear when each party is asked to define their own boundaries. Disputes are mainly related to tenure, which 'determines who can (and can't) do what with the property in question and under which circumstances they can (or can't) do it' (Lynch and Alcorn 1994: 373–4). Property is defined as 'a bundle of rights' (Bruce 1998: 1) and responsibilities (Lynch and Alcorn 1994: 374), which can be held by a state, a corporation, an organisation, a family, an individual or a community. These rights, which are complex and often overlap, have spatial, temporal, demographic and legal dimensions.

In Indonesia, conflict over land usually arises between indigenous communities and the state (Ruwiastuti 1997: 55) because state-created property rights overlap with customary (*adat*) rights. This is often the case when conflict arises between the holders of timber concessions and members of indigenous communities. Timber concession holders use state forestry laws and maps to define and claim their rights, while indigenous communities claim that customary (*adat*) rights entitle them to stake ownership over the land that their ancestors have long lived on. Similar conflicts can also arise over protected forest areas and land designated for large-scale development activities such as open-cut

[1] This chapter is based on a research project undertaken for a Masters in Environmental Management and Development at the Australian National University. The author would like to thank his two supervisors, Dr Padma Lal and the late Dr Elspeth Young, for their input to this project, as well as Judy Bell, who helped with corrections to the project report. Jaringan Kerja Pemetaan Partisipatif provided financial support for fieldwork in East Kalimantan from 26 June to 25 July 2000, with funding derived from DFID's Multi-Stakeholder Forestry Programme based in Jakarta. Additional support was provided by Yayasan Plasma, the USAID-funded Natural Resource Management Project (Kaltim), and Lembaga Bina Benua Puti Jaji, which organised trips to several villages, provided facilities and covered costs during fieldwork. Individual debts are owed numerous people involved in group discussions and interviews while collecting information for the case study: Cristina Eghenter, Andris Salu, Dolop Mamung, Phantom, Niel Makinuddin, Sulaiman Sembiring, Ade Cahyat, Miriam, Fajar, Graham Usher, David Craven, Paulus Kadok, Ana, Bonifasius Juk, and many people in the villages where fieldwork was undertaken. Special thanks are due to local government staff of Kutai Barat district, to Syahruddin who assisted with fieldwork in Kutai Barat, and to Paulus Kadok and Elisabeth who helped with trips to Sungai Belayan. Eva Gastener, Longgena Ginting, Restu Achmaliadi and Serge Marti provided additional literature and data, and Anne Casson helped with the final drafting of this chapter.

mines, transmigrant settlements and plantations. A lack of understanding and recognition of indigenous customary laws and practice (*hukum adat* and *hak ulayat)* are major factors in these land use conflicts (Peluso 1995: 391; Ngo 1996: 137).

The Indonesian government has long been criticised for managing natural resources poorly within the Indonesian archipelago. During the Suharto era, Indonesia lost over 20 million hectares of forest between 1985 and 1997 (Holmes 2000: 3) and another 10 million hectares of agricultural and forest land was burned during the 1997–98 forest fires (McCarthy 2000: 91). Commercial interests, producing 11.5 million tons of palm oil in 2004 (USDA 2005), have also contributed to unprecedented forest conversion in Sumatra and Kalimantan. In addition, because of the anticipated timber shortage and the need to decrease the exploitation of natural forest, Industrial Timber Plantations (Hutan Tanaman Industri) have been promoted by the government (McCarthy 2000: 114–15). Use of a monoculture of fast-growing species in these estates has changed the microclimate and increased the risk of large-scale fires.

Indigenous communities are often marginalised by these large-scale development activities (de Jong 1997: 188). This is because most of their *adat* lands overlap with industrial timber estates and oil palm plantations, and the government has categorised these lands as grasslands or unproductive lands to be converted into productive uses. This has led to increasing calls for land reform and more sustainable resource-management options, such as involving indigenous communities in land use decisions and allowing them to incorporate their own approaches to natural resource management into a system of community-based management.

In response to land use conflicts on the ground and the demand for equity in accessing land and resources, some research institutions and non-government organisations (NGOs) have worked together with indigenous communities to use maps as a tool for identifying and obtaining formal recognition of indigenous rights to land and natural resources. This has led to community mapping — termed 'counter-mapping' by Peluso (1995) because it takes a bottom-up approach. In order that alternative management systems for natural resources can be proposed, these maps are being used to document indigenous management systems (Peluso 1995; Stockdale and Ambrose 1996).

Peluso (1995) and Sirait (1997) have identified some of the key issues underlying community mapping. On the positive side, it can empower local people and allow them to gain land rights. However, on the negative side, community mapping can freeze property rights and create a static situation for local communities. Therefore, the role of these mapping activities in reducing conflicts over land and promoting indigenous systems in the management of natural resources is ambiguous. This chapter explores this dichotomy and proposes ways in which community mapping can result in more positive outcomes.

Land Tenure and Natural Resource Conflict in Indonesia

Land tenure arrangements have undoubtedly influenced the way in which natural resources are controlled by the state and indigenous communities in Indonesia. They reflect the imposition of Western tenure systems on existing customary systems. In many cases, these arrangements replace the diverse and complex tenure systems used by local communities with a unified and simplified framework developed by the Dutch. Conflict over land or natural resources has increased as a consequence of the contradiction between these arrangements. The following sections describe state land-tenure systems in Indonesia and indigenous customary land-tenure systems in Kalimantan to shed further light on this issue.

State-Imposed Tenure Systems in Indonesia

Even though the state did not formally own all of the 'free' land, the notion of state-controlled land was interpreted, during the Suharto period, as an exclusive authority over any territories classified as *kawasan hutan* (forest area) — including all aspects of human activities within it (McCarthy 2000: 93). In other words, the state had an authority to divide forest areas into several land use categories with different policy objectives, such as timber production and conversion of the forest area into agricultural land, using the *Basic Forestry Law* (No. 5/1967) as a legal framework. As a result, a *Forest Land Use Consensus Plan* (*Tata Guna Hutan Kesepakatan*) was established in 1982. This land use plan classified 75 per cent (or 144 million hectares) of Indonesia's land as forest areas (Evers 1995: 6), and still wields influence over the planning process for such areas, although the *Land Use Management Act* (No. 24/1992) gave the National Development Planning Agency (Badan Pembangunan dan Perencanaan Nasional or BAPPENAS), the Ministry of Home Affairs and the Ministry of Environment (Kementrian Lingkungan Hidup) more possibilities to play a key role in spatial planning (McCarthy 2000: 94–5). During this period, *adat* and *hak ulayat* were not fully recognised or understood, especially in the outer islands of Indonesia (outside Java and Bali).

When Suharto resigned in mid-1998, the Habibie government was forced to address problems arising from the *Basic Forestry Law* of 1967, and a new *Basic Forestry Law* (No. 41/1999) was released in late 1999. However, while this law recognises and understands *adat* and *hak ulayat*, it only provides possibilities for the *adat* community to manage and use *adat* forest 'as long as they are evidently in place and their presence is acknowledged' (Article 67). In other words, the *adat* community can only obtain rights to use and manage *adat* land or forest if the state acknowledges their existence. They are not able to own land.

Moreover, Article 5 of the new *Basic Forestry Law* states that the Indonesian state will only recognise community rights to forest land if it can be proven that:

- the *adat* community in question is still in a group form (*paguyuban* or *rechtsgemeenschap*) and live in their own *adat* area;
- the *adat* community still follow their *adat* institutions;
- the *adat* community forest area has clear boundaries, approved and acknowledged by their neighbours;
- there is an *adat* law framework related to forest that is still practised; and
- the *adat* community still relies on the forest for subsistence, religion and social activities based on *adat* rule.

While this new regulation may give some new opportunities to *adat* communities, a management plan for *adat* forest has to be approved by the Ministry of Forestry (Article 10) and the plan must consider existing land use planning determined by the *Regional Land Use Plan* (*Rencana Tata Ruang Wilayah*).

In other words, the Indonesian state only acknowledges the rights of *adat* communities in principle rather than in practice. In principle, all forest area is controlled directly by the state framework, which gives the *adat* community the right to use and manage their *adat* forest area, but not to own it. However, the *Basic Agrarian Law* states that existing *hak ulayat* cannot be acknowledged as 'land controlled directly by the State' (Evers 1995: 5). *Adat* rights are not, therefore, explicitly clear in forest law, although they have been clarified further in Regulation No. 5/1999, which provides guidelines on how to solve problems related to the *hak ulayat* of *adat* communities. This attention to the *adat* community seems to be compatible with the idea of regional autonomy at the district level, which is governed by Law No. 22/1999 and Law No. 25/1999, and allows district governments to secure revenues from their own natural resource base.

Nevertheless, the new *Basic Forestry Law* gives *adat* communities some recognition of their rights to land and natural resources. Therefore, there is an opportunity for community mapping to play a crucial role in helping indigenous or *adat* communities to document their *adat* area, including the rights that are attached to it, and to help them create *adat* management plans to promote their own community-based natural resource management.

Adat Tenure Changes in East Kalimantan

East Kalimantan is one of the richest natural resource provinces in Indonesia. The province, which has a population of around two million, covers 211 440 square kilometres or 10.55 per cent of the Indonesian land area (Safitri et al.

1997: 26). The diversity of ethnic groups and sub-ethnic groups that live in this area reflects the diversity of resource control and tenure systems. Generally, within a community, resources concentrated in a particular area (such as bird nest caves) can be considered private property. Some wild resources, such as rattan, are also domesticated and planted by villagers in areas where it is abundant (see Eghenter, this volume). On the other hand, scattered resources, such as *gaharu* (agar wood) tend to become common property because it is difficult to privatise these resources or allocate them to individuals (Momberg et al. 1997: 170).

Most *adat* communities in East Kalimantan have formal control over territorial claims to forest areas, which have been marked geographically on natural features such as mountain ranges and rivers by past warfare or negotiations among different tribal groups (Fox 1993: 306; Momberg et al. 1997: 170). This control has long been governed by customary law (*hukum adat*) — a web of access rules which govern the use, exploitation and conversion of particular forest products (Fox 1993: 305). The largest territory covers 'continuous villages' with the same language, and the second largest territory covers three or four villages using a 'lieutenant customary law' (*temenggung adat*). Although villages and tribal groups are diverse, they share common land and tree tenure systems. Outsiders have to apply for permission to access these areas or resources. Sanctions are also applied as a form of customary law, or *adat* fines apply if violations occur in relation to resource use (Momberg et al. 1996: 6). These *adat* communities usually practize rotational swidden cultivation and harvest timber and non-timber forest products using their *adat* management systems. These systems may differ from one *adat* community to another but, in general, *adat* communities have traditionally used their local knowledge of ecosystems and soil properties to manage natural resources (Sorensen 1997: 247).

In recent years, a range of internal and external pressures has weakened indigenous tenure systems. The weakening of cultural, social and family ties is usually a response to external pressures. Less cohesion and social control within communities causes 'individualisation of communal rights'. The absorption of communal rights within an *adat* community creates a situation where outsiders, including government, have unlimited access to *adat* land for agriculture, mining, logging, road construction and other 'land hungry' development activities. Conflicts arise between these large-scale developments and local people because the state has failed to acknowledge *adat* rights when allocating concessions and development permits. Moreover, *adat* communal lands have been threatened by 'unofficial' encroachment, such as illegal land purchases and illegal logging, which are often supported by police, armed forces or local government staff (Evers 1995: 12; Eghenter 2000a).

Sometimes, *adat* institutions also break down when community members seek to gain quick profits from particular resources, such as agar wood, rattan or timber (Sorensen 1997: 249). In the past, noble families in communities with social stratification, like the Kenyah for example, more or less willingly devolved their lands to the larger community. However, a desire to accumulate wealth and engage with the modern world has driven many of these élites to exploit natural resources for personal gain. For instance, after the fall of Suharto many élites benefited from timber harvesting after Permits to Use and Harvest Timber (Ijin Pemanfaatan dan Pemungutan Kayu) based on Forest Product Harvesting Rights (Hak Pemungutan Hasil Hutan or HPHH) were allocated to individuals, primarily members of the élite within a given community. This is despite the fact that legislation governing these permits (Regulation No. 6/1999) stipulated that these rights should be allocated to *adat* communities through cooperatives.[2] Dramatic environmental change resulting from natural disasters, such as the 1997–98 forest fires, has also threatened *adat* resource management systems as these tend to break down when the resources become scarce and more valuable.

Community Mapping and Its Implications

Efforts to reduce conflict over land ownership and resource management have increased since the fall of Suharto because the state has become more willing to acknowledge indigenous rights in an era of social reform and decentralisation (Persoon and Est 1999: 1). In light of these changes, community mapping has been used as a tool to attempt to solve conflicts over land ownership and natural resource management.

The Purpose and Role of Community-Mapping Initiatives

Community mapping can be used to collect information about traditional land uses and village boundaries, and also as a tool for local decision making and conflict resolution between villages (Momberg et al. 1996). However, the use of maps for securing rights and recognition of indigenous tenurial systems is often criticised because the outcomes may not align with existing property rights and it may not be possible to establish boundaries that reflect the nature of the community (Sutton 1995). Wood (1993: 32) has argued that maps cannot 'grow or develop', but mapping or map making do. By this he means that maps show fixed boundaries but the process of mapping and the people who create the maps are dynamic. This argument also suggests that mapping might curtail property rights.

Traditional tenurial rights and the adoption of indigenous management systems have recently received considerable attention in conservation literature

---

[2] This regulation led to rapid deforestation in several areas and was officially revoked in 2002. Nevertheless, district officials continue to issue small-scale concessions in defiance of the central government.

(Kleymeyer 1994; Lynch and Alcorn 1994). In Indonesia, some case studies and projects have involved local communities in park management — for example, in the Cyclops Mountains (Mitchell et al. 1990), Kayan Mentarang National Park (Sirait et al. 1994; Sorensen 1997) and Wasur National Park (Craven 1993). Most attention has focused on the safeguarding and promotion of indigenous rights in conservation areas and community participation in co-management of the conservation area. In these processes, indigenous management systems have been recognised and integrated with conservation purposes.

Several other countries have also adopted mapping initiatives in various projects and programs. These projects are concerned with the management and co-management of natural resources and contribute to national and global environmental protection initiatives (ECOSOC 1999). For example, in Belize, the Toledo Maya Cultural Council and the Toledo Alcaldes' Association produced documents (including maps) of the Mopan and Ke'kchi Maya people's land in 1998. These provided information on the traditional and current use of their land. They include a specific description of Maya culture, land tenure, history and socio-economic activities (ibid.). A land assessment project on Aboriginal land in central Australia, which has been developed by the Central Land Council, is another example of a community mapping initiative documenting indigenous land-use planning and management (CLC 1994).

The relationship of community mapping to the nature of the community and state land use is shown in Figure 5.1, which summarises the purpose of community mapping initiatives in general. This figure shows the role that community mapping can play in helping to identify indigenous rights or boundaries, and in promoting indigenous resource-management systems. It shows that conflict over land generally arises when governments apply land use plans and boundaries which are designed without consulting the community (see the arrow connecting Parts A and C). This conflict is caused by the occurrence of two different tenure systems, namely state tenure and customary tenure.

Tenurial conflicts between the state and indigenous communities have occurred through legal disputes and land use boundary disputes. As shown in Part A of Figure 5.1, state agrarian law and forestry law delineate one type of land use; while community uses, customary law and practices delineate other land uses. Figure 5.1 illustrates a situation where village territory overlaps with state land use assigned by government in categories such as nature reserves, protected forest, timber concessions and limited production forest for selective logging purposes. Community mapping is proposed as a tool which can be used to give indigenous communities the opportunity to identify their indigenous rights over 'state' land. It may also be used to promote indigenous management systems for community-based management. This is shown in Part B of Figure 5.1.

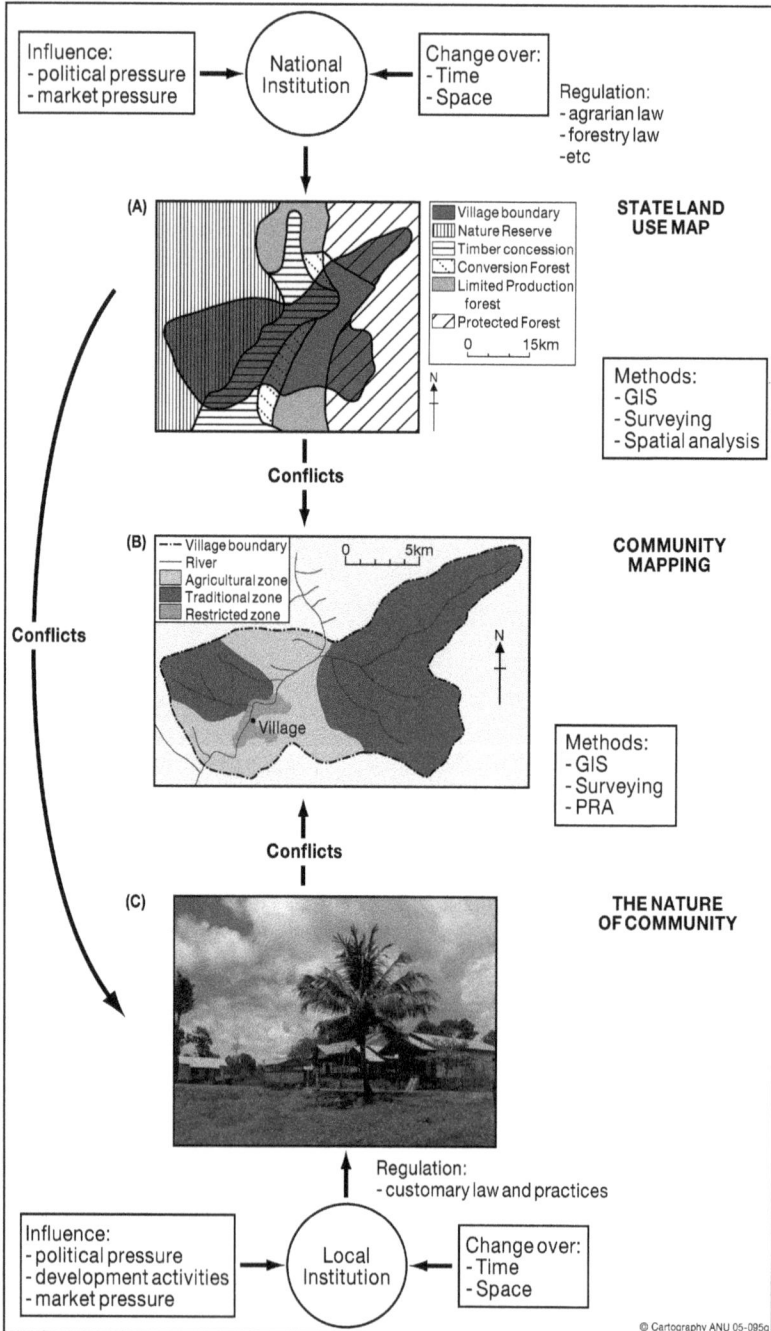

Figure 5.1. Community mapping, state mapping and the nature of community

Community mapping can allow indigenous communities to conduct local decision making and resolve conflict using spatial illustration. It can also be used as a tool to suggest alternative management strategies for natural resources. As the example in Part B of Figure 5.1 shows, village land use can fall into three zones — agricultural, traditional, and restricted. The agricultural zone is set aside for settlement and cultivation. The traditional zone is set aside for protected forest where hunting activities, timber harvesting and the collection of non-timber forest products are highly restricted. This area can be opened up for collection purposes but cultivation is prohibited. The restricted zone is used for daily subsistence purposes, such as fishing, hunting and collection of timber and non-timber forest products. A collective permit will be needed if a community member wants to open up this area for cultivation.

Conflicts of interest within the community and with other community groups often happen during the mapping process. These conflicts are sometimes difficult to avoid and to solve. This is shown by the arrow connecting Parts B and C of Figure 5.1. These conflicts will also occur when community maps are combined with state land use maps. The arrow connecting Parts A and B of Figure 5.1 describes this situation. Therefore, negotiation would be needed to combine the community's identified zones with those identified by the state.

Participatory rural-appraisal methods and spatial information technology or geomatics, such as geographic information systems (GIS), combined with other surveying technologies, are used to support community mapping initiatives and to integrate them with other information. Various governments have also used geomatics for spatial information management. The design for land use is usually based on 'scientific' criteria and the results are obtained using spatial analysis methods.

Part C of Figure 5.1 shows how the appearance of the community will change over time because of the influence of external and internal factors such as political pressure, development activities and market pressure. This is reflected by changes in the customary institutions for managing natural resources. These changes vary across space as the influences also vary. National institutions also change when domestic and international markets and political situations are involved.

Community-Mapping Activities in East Kalimantan

Demand for community mapping in Indonesia has increased over the last decade. Between 1994 and 1999, more than 1.6 million hectares of indigenous land was mapped in 14 provinces (Figure 5.2). This included 350 000 hectares (between 130 indigenous areas) in West Kalimantan (Nazarius 2000) and more than 100 village territories in East Kalimantan. Non-government organisations, research institutions and academics have played an important role in facilitating this process.

Although all of these community-mapping activities had similar concerns, different strategies were used and different backgrounds and objectives influenced their outcomes. In the case of East Kalimantan, these outcomes can be categorised into three types of initiative: those for protected area management, those with research objectives, and those for recognising indigenous rights.

Figure 5.2. Community mapping activities in Indonesia

Initiatives for Protected Area Management

The first community-mapping pilot project in East Kalimantan was conducted by WWF Indonesia in the Kayan Mentarang National Park in 1992. The project used geomatics and community-mapping techniques to assess the position and nature of forest tenure boundaries in Long Uli (Sirait et al. 1994: 411). Another community-mapping exercise was undertaken in four villages in 1994: Long Alango, Long Pujungan, Lembudud and Tang Laan. These case studies aimed to determine community perspectives about the decision to change Kayan Mentarang's status from a Strict Nature Reserve to a National Park. The research also aimed to establish a new model of community-based zoning processes (Stockdale and Ambrose 1996: 183) and has shown that indigenous land rights and resource management systems can be recognised and accommodated within a National Park (Eghenter 2000b: 1).

After Kayan Mentarang was declared a National Park in 1997, community mapping was also used to identify and resolve boundary disputes between different stakeholders and to facilitate community participation in the management of a significant conservation area (Eghenter 2000b: 4). Maps produced from these exercises were used to establish the park's overall boundaries and those of zones inside the park proposed by local communities. In addition, the maps were expected to facilitate acknowledgment of customary land and recognition of indigenous management systems (ibid.: 1).

As a part of a community development program on sustainable economic options and local capacity building, the maps were presented to local government officials and other communities for official acknowledgment (ibid.: 12). Several workshops were conducted at the village and sub-district level, before the maps were presented to officials at the district level, to reduce the potential for conflicts to arise over boundary delineations and to ensure that community members regarded the maps as legitimate. By 1998, 65 villages with approximately 1.5 million hectares of territory had been mapped within and around the Kayan Mentarang National Park (Damus 2000). Two kinds of map were produced, one showing land use and the other showing natural resource distribution. Copies were held by village leaders and by the WWF office in East Kalimantan. There were no clear rules about who could use and control the maps. The WWF combined these maps with other spatial information relating to vegetation types, animal habitats, geology, and government land-use plans to facilitate a possible consensus between the community and other parties to establish appropriate zones within the park. However, the community used the maps as a tool to negotiate their rights with other parties. For example, in 1997 *adat* leaders from the villages of Pujungan and Ulu Bahau used the maps to negotiate with the operator of a local timber concession — PT Sarana Trirasa Bakti. The maps indicated restricted forest areas (*tana ulen*) and the community leaders were able to negotiate that these areas should not be logged. They also used the maps to obtain assistance from the Minister of Forestry to help them solve this problem. In this case, the community succeeded in forcing the timber company to acknowledge their rights and traditions.[3] The Minister of Forestry encouraged the company to solve the issue and to help the villages with rural development programs.

Initiatives for Research Objectives

Since 1999, research institutes such as the Centre for International Forestry Research (CIFOR) have used community mapping techniques to document and facilitate local conflict-resolution mechanisms. The CIFOR primarily became involved in this initiative because it sought to document local conflict resolution mechanisms and to involve the local community in the regional land-use planning process. According to the *Land Use Delineation Law* (No. 24/1992) and Regulation No. 69/1996, indigenous people have a right to be involved in mapping activities. That is why CIFOR personnel used community mapping as a tool to apply a bottom-up approach to land use planning and to resolve boundary conflicts between villages on the upper Malinau River in 1999.

---

[3] During the community mapping activities in the Kayan Mentarang National Park (from 1993 to 1998), the author was involved in providing base maps, setting up the process, conducting training, and helping to compile and analyse the community maps using a GIS.

Twenty-four of the 27 villages on the upper Malinau River were mapped using community mapping techniques and most boundary conflicts were resolved during the process (personal communication, Miriam and Fajar, July 2000). The approach used in this mapping process was similar to the WWF's approach in Kayan Mentarang National Park, but in order to simplify the process, boundary mapping was undertaken before other types of mapping. Most of the facilitators were former WWF staff who had previously participated in the community mapping exercise in Kayan Mentarang National Park. Geomatic tools such as a GPS and GIS were also used to improve the map products.

Initiatives for Recognising Indigenous Rights

Indonesian NGOs have also used community mapping as a tool to obtain recognition of indigenous rights to land, forests and other natural resources in East Kalimantan. After a community-mapping network (Jaringan Kerja Pemetaan Partisipatif) was established in 1995, the improvement of human resources in facilitating community-mapping, and financing aid within local NGOs was prioritised. These NGOs have committed themselves to facilitating community mapping exercises provided that the initiative comes from the community. The mapping is meant to encourage local potential, facilitate learning, and promote equity and respect, while the mapping results become the property of the community in question (personal communication, K. Romadan, 2000).

These activities involved participatory rural-appraisal techniques in combination with traverse surveying using compasses, tapes, a GPS and a GIS. Five GIS centres were established in Indonesia to support indigenous communities in conducting their community-mapping activities. Using this technology, the information was spread among NGOs and the indigenous community.

As a result, by 1999 local NGOs together with local communities had mapped 26 *adat* areas in East Kalimantan. SHK Kaltim mapped the areas of three *adat* communities (Engkuni Pawek, Benung Pituq and Tepulang) in 1996–97 (Nazir 2000). Another local NGO called Puti Jaji facilitated community mapping in six villages (Telivaq, Mamahak Teboq, Lutan, Ujoh Bilang, Mamahak Besar and Tanjung Jaatn) and the Sungai Belayan area in 1996 (Juk 2000). Yayasan Plasma, an NGO based in Samarinda, facilitated community mapping in Gunung Menaliq, Mejaun and Lotaq in the same year, and also facilitated the mapping of Paking, Bintuan, Birun, Long Iman and Sebaing on the Mentarang River (Romodan 2000). Yayasan Padi Indonesia facilitated the community mapping of Kampong Muluy, Muara Payang, Rantau Layung, Paser Mayang, Olong Gelang, Sungai Terik, Biu, Samurangau, Simpang and Lembok (Amin 2000). Land use, settlement and regional maps were generally produced, although SHK Kaltim also produced maps showing ownership, cultural areas, natural resource distribution and areas affected by fire. These maps were primarily stored in the homes of elected community leaders and the offices of local NGOs.

Community Mapping as a Tool to Reduce Conflict over Land

Community mapping has been used as an effective tool to reduce conflict over land in East Kalimantan. For example, village boundaries have been established within and around the Kayan Mentarang National Park as a result of the community mapping initiatives undertaken in the area. A decision was also made to establish the outer boundaries of the park, which excluded the current and future areas of use by the community. In addition, core zones, traditional use zones and other zone boundaries were identified during the community mapping process and proposed to the park authorities and other parties.

The CIFOR has demonstrated that land conflicts can be solved through community mapping exercises (personal communication, Miriam and Fajar, July 2000). Through this initiative, most village boundaries along the upper Malinau River were mapped and many boundary conflicts were resolved during the mapping process. In addition, some customary (*adat*) boundaries were mapped in the districts of Kutai Barat, Kutai Induk and Pasir. Most of the villagers interviewed by the author said that they were satisfied with the boundaries that were drawn on the maps, and after seeing clearly defined boundaries they felt secure enough to enforce their own communal rights. This demonstrates that community mapping can help to define or 'formalise' undocumented customary or village boundaries.

However, in many cases, conflicts over land and other natural resources have not been resolved, and in some cases further disputes have even arisen. The disputes tended to arise because of outstanding conflict over:

- ancestral and administrative boundaries;
- vested interests driving the mapping process; and
- the current needs of the community.

These issues are significant and are discussed in the following section, which draws upon the experience of community-mapping exercises conducted by two local NGOs (Yayasan Plasma and Lembaga Bina Benua Puti Jaji) in four villages in East Kalimantan — Lotaq, Mejaun, Ritan Baru and Buluksen. Lotaq and Mejaun are located in the upper reaches of the Lawa River within Kutai Barat district, around 250 km from Samarinda. Buluksen and Ritan Baru are located in the upper reaches of the Belayan River within Kutai Tengah district, around 260 km from Samarinda (Figure 5.3). In-depth observation, using a combination of group discussions and semi-structured interviews, was conducted between 26 June and 23 July 2000 within these four villages. A combination of group discussions and semi-structured interviews was also conducted with other communities, government staff and various community-mapping facilitators, such as local and international NGOs, research centres and inter-governmental projects.

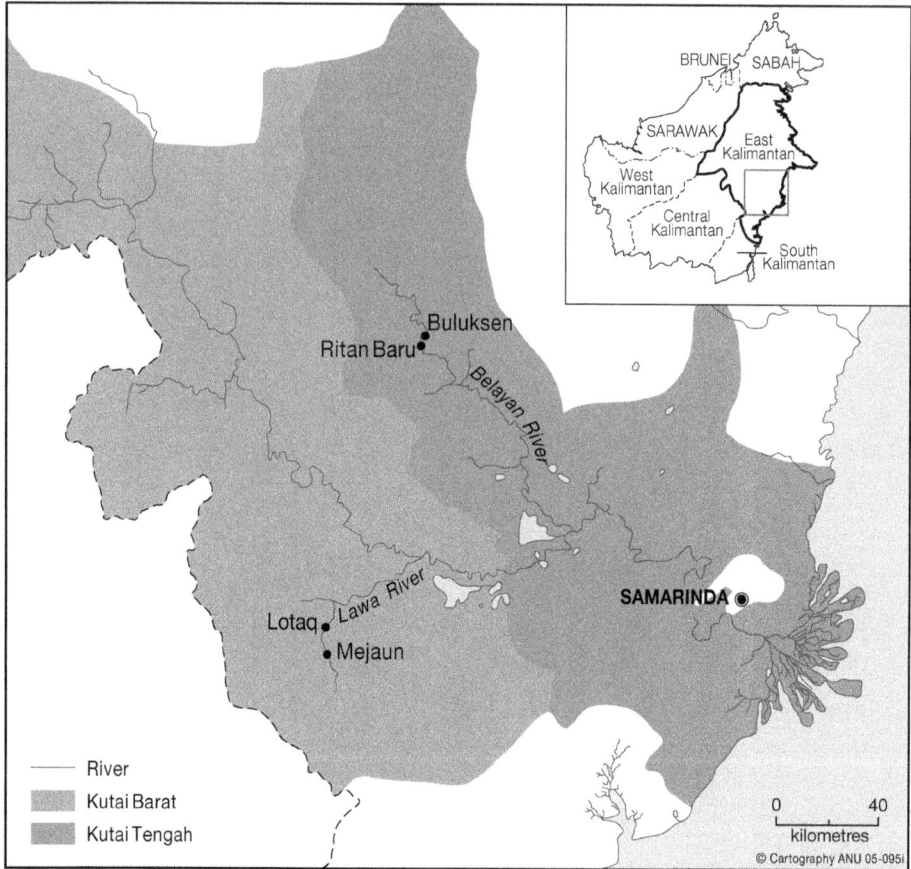

Figure 5.3. Case study locations

Ancestral and Administrative Boundaries

The debate about ancestral (*adat*) and village administration boundaries is related to the debate about the authority of both *adat* and government institutions. In some villages these boundaries are similar (personal communication, D. Amin, July 2000), but there are also cases where several villages exist within a single *adat* territory (personal communication, I. Damus and B. Juk, July 2000). Although conflict may have existed before the community-mapping process was conducted, the process has tended to clarify or formalise divisions of the *adat* territory and, as a consequence, village disputes have increased due to an increase in apprehension about the loss of private ownership.

For example, Marhum Pemarangan, the King of Kutai Karta Negara (1730–32), gave the villages of Buluksen, Ritan Baru and Long Lalang an *adat* area larger than the current mapped area (personal communication, Ana, July 2000). Administrative boundaries overlap with the *adat* area and divide it into village territories. Restricted communal property such as *tana ulen* and *tana saru* within this *adat* community is located in Ritan Baru territory, and this makes the ownership of these areas unclear. Some questions regarding communal ownership can be raised, such as: which institution will govern (control and access) the use of the former communal natural resources? Will it fall within the jurisdiction of *adat* institutions or village administrations? And who will benefit from this situation? Community mapping was conducted within these five villages in 1998. These administrative boundaries have been used as a reference for conflict resolution and the community did not have a problem with these boundaries, but members were concerned that the remaining land or resources should be protected.

The administrative boundary is not actually clear, and most of the community mapping conducted in these villages primarily helped the district government to delineate village boundaries. The choice of mapping unit, whether administrative or ancestral, raises further questions. If the administrative political unit was chosen, was it an indication of the irrelevance of indigenous management systems? If customary units were used, how could the process be implemented in an increasingly heterogeneous and commercially oriented community?

Some community members, who were not involved in the community mapping process, used the maps to propose HPHH rights (rights to harvest forest products) for their own benefit (personal communication, Suto, July 2000). It was apparent that the conversion of communal property into private property accelerated after the mapping process.

Besides the division of customary land into village territories, the administrative-boundary approach cut social ties within communities. For example, conflict between communities (Lotaq and Muara Begai) increased after the Lotaq village people made the maps. Members of the Muara Begai community were concerned that their rights to use land or resources within the Lotaq area may have been affected. This was despite the fact that the *adat* leader of Lotaq village stated that the Muara Begai community would still be able to practice swidden cultivation within the Lotaq area as long as they reported their activities to *adat* or village leaders (personal communication, July 2000). However, the discussion proved that no clear regulation had been approved to address this problem during the mapping process.

The above examples illustrate that the complexity of indigenous tenure, which has governed the use of land or resources and governed the ownership of resources such as trees as well as the social relationship between villages, has been simplified or frozen by 'clear' boundary regimes produced by some of the community-mapping processes. Atok (1998: 46) stated that common discussions about ancestral or *adat* boundaries centred around different concepts of boundaries. *Adat* boundaries do not usually form neat lines as they tend to follow natural features, such as rivers or mountain ranges. This needs to be considered before conducting mapping exercises. Empowerment of *adat* institutions to enforce their regulations, understanding local conditions and indigenous tenure systems including the boundary concept, and considering the impact of map production on the indigenous tenure system, have to be taken into account before deciding which unit of mapping (administrative or ancestral) will be used. The community-mapping process, where all parties sit together to solve their problem using spatial tools as a medium of discussion, can play a key role in resolving conflicts over land and resources provided it does not encourage a set of boundaries on maps that ignore the indigenous tenure system.

Vested Interests behind Community Mapping

In addition to the potential for conflict to arise over *adat* or administrative boundaries, an assessment of mapping activities in these four villages also demonstrated that there are different interests involved in boundary claims. Most of these interests have been driven by a desire to exploit natural resources. For example, conflicts between Lotaq and Muara Begai villages have been steered by interests in coal deposits within both areas. When fieldwork for this study was completed in 2000, no consensus or decision had been reached about how to solve this boundary dispute. Lotaq villagers wanted to enforce their ancestral boundaries, but Muara Begai villagers disputed the village boundaries mapped by the Lotaq community. According to the *adat* leader of Lotaq village, the Muara Begai community had claimed some coal deposits within the Lotaq area (personal communication, Ahen, July 2000). Some of the Muara Begai community representatives attended the mapping process in Lotaq village but they did not complain about the boundaries during this process. The dispute only started when the Lotaq villagers later asked for their maps to be approved by the Muara Begai community.

Representation and responsibility are fundamental issues in a participatory community-mapping process. Most of the respondents interviewed did not know about the mapping process and were not directly involved in it. For example, in the village of Lotaq, women rarely participated in the mapping process because they were kept busy preparing food for those attending the meetings. Field survey work was generally conducted by younger men. However, during my own fieldwork, *adat* and village leaders said that all of the community members were involved in the mapping process. This information was contradicted when the same question was put to some women and younger men, who responded that they came to the place where the sketch map was produced but did not participate in the technical mapping. Nevertheless, my interviews suggest that most of the people within the Lotaq community were satisfied with the result because they thought the maps would protect their territory from encroachment.

A different situation was found in the village of Buluksen. None of the respondents from this village knew about the community-mapping process because all of the people involved in the process were pursuing their own interests. In this particular case, two local NGOs facilitated community mapping in five villages along the Belayan River to protect village territory from logging companies and to promote indigenous forest management systems. The Buluksen community was only represented by village officials. It was later revealed that these officials then went to Samarinda to organise permits for harvesting forest products for their own benefit. These members of the village élite were driven by a desire to use common agreements for their own profit. Maps produced through the mapping process were used as evidence of agreement within the community about land use — a prerequisite for obtaining HPHH. This situation increased conflict within the Buluksen community. Many community members did not want to attend village meetings unless these were facilitated by outsiders such as NGO staff. A similar situation also occurred in Ritan Baru, where competition arose between members of the élite after they sought to obtain private rights over a communal forest using maps as evidence of communal interests.

To make sure that maps will be used for communal purposes, almost all of the villagers involved in the community-mapping process stated that those who want to use the community maps should consult with the *adat* or village leader and have their request approved by all community members.[4] However, it was not clear how the community itself should control or use maps stored in the *adat* or village leader's house (personal communication, Ahen, July 2000). Outsiders could still access the maps without obtaining permission from all community members. For example, in the village of Lotaq, agrarian staff were able to obtain these maps from the village leader's wife and the maps were not returned. The community was hence worried that unauthorised parties would use the maps for their own purposes and this threatened the community's rights over the land.

Government interest in community mapping has recently increased. This may be driven by a desire to find solutions to conflicts between local communities and logging companies, but could also be driven by a search for 'empty' land (*tanah kosong*) by investors. The idea of an 'empty forest' reflects a lack of understanding about indigenous tenure systems, since the 'empty' space drawn on the maps does not mean 'empty' in real terms. The interest in maps produced through community mapping can divert attention away from gaining an

---

[4] From a group discussion in Lotaq, 6 July 2000.

understanding of the complex nature of indigenous communities and their rights to 'empty' forest. As a consequence, community mapping can be used as a cheap means of data collection for government planners.

Indigenous Communities in Transition

Conflicts over the management of natural resources have resulted from various external and internal changes. Migration, resettlement and regrouping of villages, as well as various development activities, have made most communities heterogeneous, with changes in their interests, knowledge and livelihoods, as well as changes to their environment. Because of these changes, conflicts will always occur.

Under these conditions, those leading community mapping processes need to endeavour to facilitate the sharing of power within communities so that local élites can be controlled and the *adat* land rights remain as communal rights rather than private property rights. This was highlighted in a group meeting in the village of Mejaun when a participant said that 'natural boundaries are fixed but people change'. Through the involvement of all parties (including migrants) in the process, community mapping can accommodate power sharing and represent the needs of all parties.

Changes have also occurred in the environment. For example, in the case of Lotaq village, almost 80 per cent of the community's forest was burnt during the 1997–98 forest fires. As a result of this loss the villagers can no longer practise their traditional systems to manage remaining natural resources. In this situation, mapping the former condition of land use can be just as important as documenting how to protect their land from outsiders.

Conclusions

Community mapping has been widely used in East Kalimantan to secure indigenous property rights and promote community-based management of natural resources. In most cases, conflicts over land and natural resources have been solved during the mapping process through delineation of boundaries. Such conflicts may occur between local communities and the government, between communities, and within communities. Conflicts between communities are especially likely in relation to claims over areas that have high economic value, such as mineral deposits or timber concessions. These disputes highlight debates over ancestral and administrative boundaries, ownership of resources and land, and the rights of other parties, including migrants. In addition, conflict among village members has increased as a result of village élites allocating natural resources for their own economic benefit. As shown in the case study, in some villages maps were used to legitimise individual ownership rather than communal village ownership.

This demonstrates that conflicts remain and the mapping process tends to be driven by élites and facilitators. Consequently, the effectiveness of community mapping in promoting community-based management of resources remains questionable. However, local economic and social development can be achieved when power sharing between parties leads to more equitable and sustainable resource use. The land-use planning process in community mapping can provide opportunities for participatory democracy and decentralised decision making where an effective conflict-resolution mechanism can be established. However, these opportunities may be compromised by the way that maps are used when the mapping process is complete.

Some conflict over natural resource management is generally unavoidable and is part of the dynamic nature of indigenous communities. However, this conflict should be managed in order to maintain stability within the community and sustain equity in resource use. The combination of co-management and adaptive management, where management adapts to changing ecological and social conditions, may enable greater involvement of indigenous communities in natural resource management. In addition, adaptive conflict management, through repetition of community-mapping processes, may be an appropriate solution for maintaining power relations and equity within communities.

References

Amin, D, 2000. 'Membuat Peta Kampong Kita untuk Generasi Anak Cucu [Let's Make a Map of our Village for our Grandchildren].' Paper presented at a workshop on 'Participatory Mapping and its Influence in Regional Planning', Samarinda, 11–12 July.

Atok, K., 1998. 'Konsep Pemetaan Partisipatif [The Principal of Participatory Mapping].' In K. Atok, P. Florus and A. Tamen (eds), *Pemberdayaan Pengelolaan Sumber Daya Alam Berbasis Masyarakat [Empowering Community Based Natural Resource Management]*. Pontianak: Pancur Kasih.

Bruce, J.W., 1998. 'Review of Tenure Terminology.' *Tenure Brief* 1: 1–8.

CLC (Central Land Council), 1994. 'The Development of a Land Resource Assessment Project on Aboriginal lands of Central Australia.' Paper presented at the 'Ecological Society of Australia Conference', Alice Springs, September 1994 (citation based on permission from Elspeth Young).

Craven, I., 1993. 'People's Participation in Wasur National Park.' *Review Indonesia* 79: 32–33.

Damus, I, 2000. 'Pemetaan Desa Partisipatif di Taman Nasional Kayan Mentarang, Kabupaten Malinau dan Nunukan, Kalimantan Timor [Participatory

Village Mapping in the Kayan Mentarang National Park, Malinau and Nunukan Districts, East Kalimantan].' Paper presented at a workshop on 'Participatory Mapping and its Influence in Regional Planning', Samarinda, 11–12 July.

de Jong, W., 1997. 'Developing Swidden Agriculture and the Threat of Biodiversity Loss.' *Agriculture, Ecosystems and Environment* 62: 187–197.

ECOSOC (UN Economic and Social Council), 1999. 'Human Rights of Indigenous Peoples: Indigenous People and their Relationship to Land.' New York: ECOSOC.

Eghenter, C., 2000a. 'Histories of Conservation or Exploitation? Case Studies from the Interior of Indonesian Borneo.' Paper presented at the conference on 'Environmental Change in Native and Colonial Histories of Borneo: Lessons from the Past, Prospects for the Future', Leiden, 10–11 August.

———, 2000b. *Mapping People's Forests: The Role of Mapping in Planning Community-Based Management of Conservation Areas in Indonesia.* Washington (DC): Biodiversity Support Program.

Evers, P.J., 1995. 'Preliminary Policy and Legal Questions about Recognizing Traditional Land in Indonesia.' *Ekonesia: A Journal of Indonesian Human Ecology* 3: 1–23.

Fox, J., 1993. 'The Tragedy of Open Access.' In H. Brookfield and Y. Byron (eds), *Southeast Asia's Environmental Future: The Search for Sustainability.* Singapore: Oxford University Press.

Holmes, D., 2000. 'Deforestation in Indonesia. A Review of the Situation in 1999.' Jakarta: draft report to World Bank.

Juk, B., 2000. 'Pemetaan Partisipatif Salah Satu Cara Melibatkan Masyarakat dalam Menata Wilayah Desa [Participatory Mapping as a Kind of System of Involving the People in the Layout of the Village Region].' Paper presented at a workshop on 'Participatory Mapping and Its Influence in Regional Planning', Samarinda, 11–12 July.

Kleymeyer, C.D., 1994. 'Cultural Traditions and Community-Based Conservation.' In D. Western and R.M. Wright (eds), 1994. *Natural Connections: Perspectives in Community-based Conservation.* Washington (DC): Island Press.

Lynch, O.J. and J.B. Alcorn, 1994. 'Tenurial Rights and Community-Based Conservation.' In D. Western and R.M. Wright (eds), op. cit.

McCarthy, J.F., 2000. 'The Changing Regime: Forest Property and *Reformasi* in Indonesia.' *Development and Change* 31: 91–129.

Mitchell, A., Y.D. Fretes and M. Poffenberger, 1990. 'Community Participation for Conservation Area Management in the Cyclops Mountains, Irian Jaya, Indonesia.' In M. Poffenberger (ed.), *Keepers of the Forest: Land Management Alternatives in Southeast Asia*. Hartford (CT): Kumarian Press.

Momberg, F., K. Deddy, T.C. Jessup and J. Fox, 1996. 'Drawing on Local Knowledge: Community Mapping as a Tool for People's Participation in Conservation Management.' Unpublished report to WWF Indonesia Program, East West Centre and Ford Foundation.

———, R. Puri and T. Jessup, 1997. 'Extractivism and Extractive Reserves in the Kayan Mentarang Nature Reserve: Is Gaharu a Sustainably Manageable Resource?' In K.W. Sorensen and B. Morris (eds), *People and Plants of Kayan Mentarang*. Jakarta: WWF Indonesia Program.

Nazarius, T.H., 2000. 'Development of Community-Based Management through Participatory Mapping by *Adat* Community in Forest Areas or Agroforestry in West Kalimantan.' Paper presented in Jakarta, 23 May.

Nazir, T., 2000. 'Using Community Mapping to Identify Customary Rights in East Kalimantan.' Paper presented at a workshop on 'Participatory Mapping and its Influence in Regional Planning', Samarinda, 11–12 July.

Ngo, T.H.G.M., 1996. 'A New Perspective on Property Rights: Examples from the Kayan of Kalimantan.' In C. Padoch and N.L. Peluso (eds), *Borneo in Transition: People, Forests, Conservation, and Development*. Oxford: Oxford University Press.

Peluso, N.L., 1995. 'Whose Woods Are These? Counter-Mapping Forest Territories in Kalimantan, Indonesia.' *Antipode* 27: 383–406.

Persoon, G.A. and D.M.E. Est, 1999. 'Co-Management of Natural Resources: The Concept and Aspects of Implementation.' Paper presented at a workshop on 'Co-Management of Natural Resources in Asia: A Comparative Perspective', Cabagan (Philippines), 16–18 September.

Romadan, K., 2000. 'Belajar dan Bekerja Bersama Masyarakat Membuat Alat Bantu: Pengalaman Fasilitasi Pemetaan Inisiatif Masyarakat [Learning and Working with the People in Building Helpful Tools: Experience in Facilitating the Mapping Initiative of the People].' Paper presented at a workshop on 'Participatory Mapping and its Influence in Regional Planning', Samarinda, 11–12 July.

Ruwiastuti, M.R., 1997. 'Hak-Hak Masyarakat Adat Dalam Politik Hukum Agraria [Indigenous Rights in Agrarian Political Law].' In D. Bachriadi, E. Faryadi and B. Setiawan (eds), *Reformasi Agraria: Perubahan Politik, Sengketa, dan Agenda Pembaharuan Agraria di Indonesia [Agrarian Reform: Political Changes, Conflicts, and Agrarian Reform Agendas in Indonesia].* Jakarta: Fakultas Ekonomi Universitas Indonesia.

Safitri, M.A., A. Kusworo and B. Philipus, 1997. *Peran dan Akses Masyarakat Lokal dalam Pengelolaan Hutan: Kajian Kebijakan Daerah Lampung, Kalimantan Timur dan Nusa Tenggara Timur [Influence and Access of Indigenous Communities in Forest Management: Policy Assessment for Lampung, East Kalimantan and Nusa Tenggara Timur].* Jakarta: Universitas Indonesia.

Sirait, M.T., S. Prasodjo, N. Podger, A. Flavelle and J. Fox, 1994. 'Mapping Customary Land in East Kalimantan, Indonesia: A Tool for Forest Management.' *Ambio* 23: 411–417.

Sirait, M.T., 1997. Simplifying Natural Resources: A Descriptive Study of Village Land Use Planning Initiatives in West Kalimantan, Indonesia. Manila: Ateneo de Manila University (M.Sc. thesis).

Sorensen, K.W., 1997. 'Traditional Management of Dipterocarp Forest: Examples of Community Forestry by Indigenous Communities with Special Emphasis on Kalimantan.' In K.W. Sorensen and B. Morris (eds), op. cit.

———— and B. Morris (eds), *People and Plants of Kayan Mentarang.* Jakarta: WWF Indonesia Program.

Stockdale, M.C. and B. Ambrose, 1996. 'Mapping and NTFP Inventory: Participatory Assessment Methods for Forest-Dwelling Communities in East Kalimantan, Indonesia.' In J. Carter (ed.), *Recent Approaches to Participatory Forest Resource Assessment.* London: Overseas Development Institute.

Sutton, P., 1995. *Country: Aboriginal Boundaries and Land Ownership in Australia.* Canberra: Aboriginal History Inc. (Monograph 3).

USDA (United States Department of Agriculture), 2005. 'Indonesia: Palm Oil Output Expansion Continues.' Washington (DC): USDA Production Estimates and Crop Assessment Division.

Western, D. and R.M. Wright (eds), 1994. *Natural Connections: Perspectives in Community-based Conservation.* Washington (DC): Island Press.

Wood, D., 1993. *The Power of Maps.* London: Routledge.

Chapter Six

Community Cooperatives, 'Illegal' Logging and Regional Autonomy in the Borderlands of West Kalimantan

Reed L. Wadley

Introduction

After the onset of the Indonesian economic crisis in 1997, 'illegal'[1] logging increased quite dramatically across the country. In West Kalimantan, these activities invariably involved the export of timber across the porous international border into Sarawak, Malaysia. (The same has held true for East Kalimantan, with timber going into Sabah.) The power vacuum left after the end of Suharto's New Order regime resulted in a *de facto* regional autonomy, well prior to the implementation of formal *otonomi daerah* in 2001 which has continued to facilitate these logging and export activities.

In the borderland of the upper Kapuas River, local élites and Malaysian timber bosses have taken advantage of this situation and of the 1999 forestry law permitting community cooperatives to cut timber for sale, creating an economic mini-boom. Many communities have become part of registered cooperatives whose ostensible aim has been community development. In practice, the goal has been logging, with the wood being transported across the international border into Malaysia. (Sawmills have been built on the Indonesian side of the border, but the lumber cut there has ended up in Malaysia.) The communities have received commissions for the timber extracted from their lands, but this has generally amounted to less than one per cent of the export value of the wood.

Occasional news reports of 'illegal' logging and smuggling of cut timber have appeared in the national and regional press, but efforts to stop it have tended to be very meagre. Locals have been of the strong opinion that nothing would be done about it because of local-level corruption, with government officials, military and police being paid off by the timber bosses or their representatives. In addition, there has been a challenge to local communities' territorial boundaries. Since this logging boom began there have been a number of instances of community disputes over forest. In at least one case, the dispute was over forest land that had never been part of any traditional community territory.

---

[1] I have deliberately placed 'illegal' in quotation marks to highlight the problematic nature of this complex phenomenon.

Locals have seen this as a rush to make claims on timber resources so that the local profits from logging might go to them.

In this chapter, I consider the question of how local communities control or influence the practice of 'illegal' logging within and across community and state boundaries. Drawing on field research from the upper Kapuas borderland in the vicinity of Danau Sentarum National Park, I examine an overlooked but important factor in this logging — the establishment of the borderland and the concomitant development of a 'borderlander' identity among the Iban inhabitants. I then look at local community cooperatives, the practice of 'illegal' logging, and the influence of regional autonomy on these activities. I also examine local perceptions of the situation, their worries about future impoverishment, and the role that local empowerment can and does play in dealing with regional and foreign interest in their forests. In addition, I explore the potential of local, low-mechanised logging for sustainable forest management.

Borderlands and Borderlanders

Boundaries separating nation-states perform various functions: for example, restricting and excluding labourers and diseases, preventing smuggling, collecting taxes and duties and defining citizenship and legal jurisdiction. The modern concept of nation-state boundaries spread virtually worldwide through European colonialism (Boggs 1940: 23–4; Asiwaju 1983: 2–3). Under this notion, borders should be precisely defined, clearly demarcated, jealously guarded, and exclusive. As a result, states see borders as lines separating distinct social systems.

Yet borders worldwide resemble one another as arbitrarily imposed lines of demarcation, often dividing similar areas and people, sometimes into mutually hostile states (Asiwaju 1983: 9–10). The regions along such boundaries are often unique social systems in themselves, defined by the movement of people, goods and ideas across the border and by the forces behind that movement (Martinez 1994; Alvarez 1995). The unity of a people within a boundary zone is thus often greater than that of the borderlanders with the heartland (Boggs 1940: 6). Indeed, 'cross-border informal linkages ... generally operate often to the embarrassment of all modern states everywhere in their inherent concern to keep their borders as clear and visible as possible' (Asiwaju 1983: 18).

Colonial and state boundaries have imposed different symbols of formal status upon the same ethnic groups, mainly in the form of citizenship. Boundaries were drawn across well-established lines of communication, including: a sense of community based on common traditions; usually very strong kinship ties; shared socio-political institutions; shared resources; and sometimes common political control. Colonisation brought different education systems and different official languages that have often persisted after independence. In many cases, ethnic groups divided by borders were given different names on either side. Yet despite the imposed boundaries and accompanying divisions, partitioned peoples in many Third World situations largely ignore the border in their daily lives (Asiwaju 1985).

Border areas are characterised by a high degree of peripherality, wherein often minority ethnic groups face disadvantage *vis-à-vis* the élites who control the state and see the borderland from a standpoint of strategic territorial advantage against potentially rival states. Borderlanders are more culturally and economically independent and less willing to adopt the national culture (Rumley and Minghi 1991), leading in some cases to attempts at secession from the state (Martinez 1994). In other cases, cross-border migrations occur, spurred by the desire for sanctuary against taxes, to escape political and economic oppression, or to take advantage of economic opportunities (Asiwaju 1976, 1983, 1985).

Borders 'invariably separate inequalities' (Asiwaju 1983: 19), so borderlanders have a casual and enterprising attitude given their need to be resourceful in exploiting changing border conditions. As mentioned above, borderlanders are often politically ambivalent (Asiwaju 1985). They try to manipulate their national identities, with many people acquiring and claiming citizenship in different countries and taking advantage of the rights and privileges of citizenship, but rarely exercising the corresponding duties (Martinez 1994: 20, 313).

Borderlanders also develop interests that may conflict with the state or national interest, fostering a high degree of alienation from the core. Borderlanders may thus find it acceptable to breach laws that they perceive as being at odds with cross-border interaction and thus their own interests, such as in smuggling. Smuggling is often fuelled by cross-border ethnic ties. The borders in many Third World countries are often not patrolled, may be impossible to patrol, are occasionally unmarked, and thus are not a barrier to trade but rather a conduit of people and goods. Indeed, what states regard as smuggling is often everyday economic activity within a group of closely related people (Asiwaju 1976).

The Upper Kapuas Borderlands

This general outline of borderlands describes quite well the historical and contemporary situation along the border separating West Kalimantan and Sarawak. Here, I focus on that part of the border inhabited by the Iban and in the vicinity of the national park (Figure 6.1).

The border between Dutch-held West Borneo and British-controlled Sarawak developed from the 1840s following the establishment of James Brooke's kingdom in Sarawak. Brooke sought to extend British influence in the western archipelago, and his presence in Sarawak led the Dutch into numerous attempts to establish

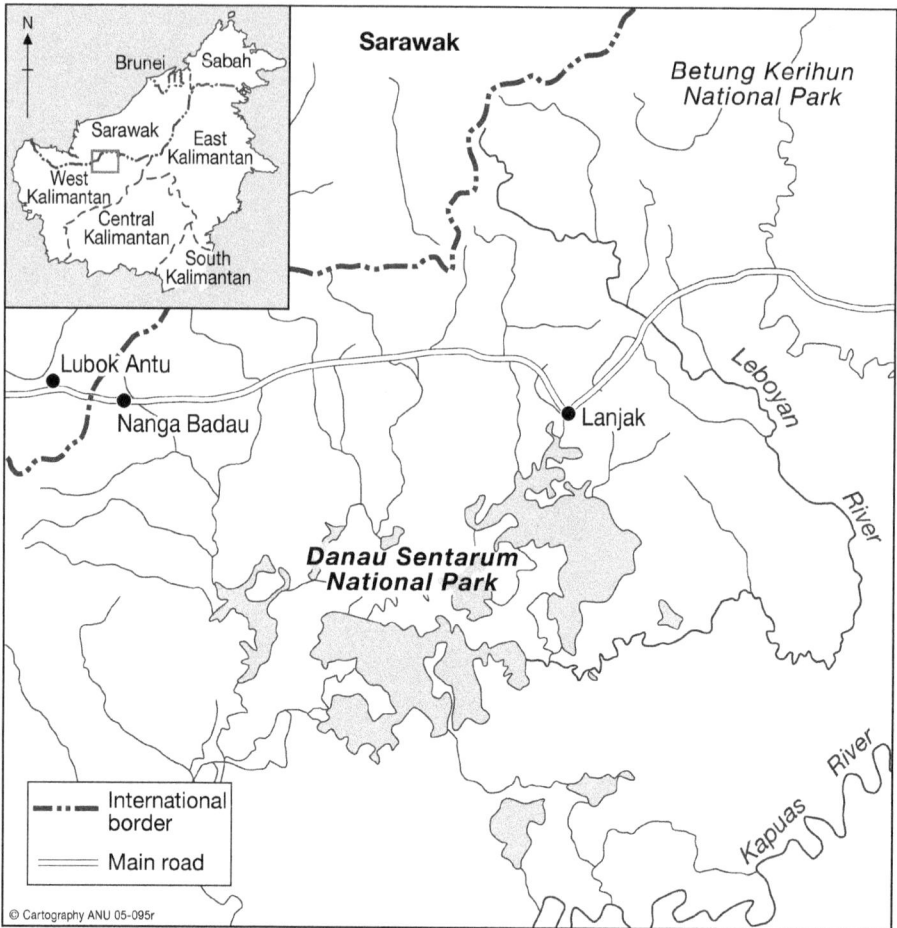

Figure 6.1. Location of the Danau Sentarum National Park

and clarify an inter-colonial border (Wadley 2001). An understanding developed between the Dutch and Sarawak governments, that the generally low-lying watershed between the north-flowing and west-flowing rivers formed the inter-colonial boundary. This demarcation, which held for several decades, effectively partitioned a number of ethnic groups inhabiting the area but was not formally set down in a treaty until 1891.

Early on, the Dutch were very concerned about the contacts James Brooke had made with the various rulers along the Kapuas. They knew Brooke was interested in stimulating trade across the frontier but also worried that he might destabilise Dutch control of the area through much lower rates of exchange for essential goods such as salt, and a lucrative trade in firearms and ammunition.[2]

---

[2] Algemeen Rijksarchief Netherlands, Geheim Verbaal 30 January 1847 No. 49, 17 July 1847 No. 255/D1, 28 September 1847 No. 335.

The Dutch observed that the Sarawak ruler was in communication with the various Kapuas rulers regarding matters of trade and disputes with Dayaks, and they worried that the border Dayaks would fall under his influence through the salt trade.[3]

The frontier between the Batang Lupar River in Sarawak and the extensive Kapuas Lakes area provided fairly easy access from the upper Kapuas to the north coast. In fact, the point where the old trail crosses the watershed (and border) is only 72 m above sea level and, in the early 1850s, the 'path lay constantly through narrow valleys, in which the ascent was very trifling' (Pfeiffer 1856: 73). In peacetime, Malay traders settled along this route by which 'some trade is carried on from Sintang and other places in the interior with Singapore… [In 1839 and prior to Brooke's arrival] a quantity of fire arms was brought that way from Singapore [to Sintang]' (Anonymous 1856: 121).

This was one area through which the Dutch suspected Brooke of seeking trade links into the Kapuas.[4] They regarded the native trade between the Kapuas and the north coast, which had probably existed for centuries, as smuggling. Dutch concern grew when Brooke established an outpost at Nanga Skrang (later Simanggang) on the Batang Lupar River where Dutch 'subjects' could buy salt and other goods at far lower prices than through sanctioned Dutch channels (Kielstra 1890: 1483–5).[5]

These trade concerns were complicated by the existence of the very large, assertive Iban population along the watershed. In 1855, Iban leaders (on the Kapuas side of the frontier) formally pledged their allegiance to the Netherlands Indies Government. They agreed to cease headhunting, to bring all disputes to the government, and stop trading in smuggled goods. The Dutch specifically forbade trade with Sarawak in salt, opium and tobacco. This 'treaty' with the Iban began a very troublesome relationship between the Dutch and the Iban, and between the Dutch and Sarawak, over frontier Iban affairs (see Kater 1883; Niclou 1887; Pringle 1970; Wadley 2000, 2001, 2003).

The decades of the 1860s–80s were the most troublesome for both the Sarawak and Dutch governments as they sought to control Iban raiding, migration and farming across the border. Foremost among these problems for the Dutch was what they saw as Sarawak's failure to respect their territorial sovereignty and control its subjects' cross-border activities. The Dutch were particularly concerned with defining Iban citizenship, and the Iban themselves continually confounded their efforts, behaving as classic borderlanders and using either

---

[3] Arsip Nasional Republik Indonesia, West Borneo Residency, No. 128, Reisverslag 1847, and No. 224, Aanteekeningen 1847; Algemeen Rijksarchief Netherlands, Openbaar Verbaal 16 March 1859 No. 30.
[4] Arsip Nasional Republik Indonesia, West Borneo Residency, No. 17, Algemeen Verbaal 1856; Algemeen Rijksarchief Netherlands, Geheim Verbaal 11 January 1856 No. 15.
[5] Arsip Nasional Republik Indonesia, West Borneo Residency, No. 58, Politiek Overzicht 1854.

side of the border to flee punishment for raiding and to escape increases in taxes (Kater 1883; Wadley 2001).

Even after Iban pacification and a lessening of Dutch–Sarawak rivalry in the early 20th century, the Iban maintained their position and identity as borderlanders. Post-independence conditions have only enhanced this (*Kompas Cyber Media Online*, 7 August 2000), especially with the consistently much stronger Malaysian economy offering opportunities to people who can easily pass as Malaysian citizens (Wadley 1997).

Indonesian involvement in this area came in 1963 when President Sukarno devised *Konfrontasi* (Confrontation) with the newly formed Federation of Malaysia, sending troops to the Sarawak–West Kalimantan border in an ostensible attempt to protect the borders from British neo-colonialism and to aid insurgents in Malaysia (Mackie 1974). When Suharto came to power in 1966 and orchestrated the massive purge against communists and suspected communists, *Konfrontasi* along the border phased into an even heavier military presence aimed at wiping out the communist insurgents who fled to the area to continue their struggle. This campaign lasted into the early 1970s and resulted in a great deal of disruption to the lives of the Iban borderlanders. Not only was there a cramp on cross-border activities such as trade and visiting kin (McKeown 1983), but for Indonesian Iban, political and economic loyalty was often in question (Soemadi 1974; Wadley 1998). This set the stage for subsequent outside resource exploitation under the New Order regime.

From the 1970s, under Suharto's national development program, the government granted huge timber concessions throughout the archipelago. Following the *Basic Forestry Law* of 1967, these concessions were laid over and largely ignored local claims to land and forest, as they operated 'in the national interest' and therefore superseded local rights. In the West Kalimantan borderlands, these concessions invariably involved Indonesian military connections derived from the earlier anti-insurgency efforts. One concession, that of PT Yamaker, combined economic exploitation with national security concerns, stretching along the border from Tanjung Datu in the west to the upper Leboyan in the east. The company leadership was largely ex-military. In addition to Yamaker in the Iban borderland, there were three other companies, one of which was controlled by local border Iban élites who made powerful allies by helping the military during the insurgency.

During the heyday of concession logging in the 1980s and early 1990s, timber companies operated with broad and state-supported authority over the forests granted to them. In dealing with local communities there was often talk about gaining permission from the locals to log, and of promises of jobs and development, but even locals were generally aware that the companies did not need permission from them if they held permits from the government.[6] Company representatives and their contractors paid honoraria to local officials for their cooperation,[7] and they made occasional, but half-hearted, gestures to aid affected communities.[8] In consultation with district *adat* leaders they also provided compensation to locals for damage to fruit and rubber trees or to forest cemeteries during road building and logging work. But the fines levied were generally 60 per cent less than those levied on locals.[9] In addition, locals tended to shun the jobs available because of the low wages offered. In the mid-1990s, the border area was facing the looming prospect of oil palm plantations and further loss of local access to old resources (Wadley et al. 2000).[10]

This situation continued until the fall of Suharto in 1998, the ensuing economic crisis (*krismon*), and the eventual election of a 'reform' government. The new government ended some concessions in the border area, including that of PT Yamaker (which was taken over by the government corporation PT Perum Perhutani III). An oil palm concern (laid over part of Yamaker's concession) escaped the axe, as did the 'Iban' concession, which was even granted a 10-year extension because its leaders argued that it was a local concern dedicated to local development (Harwell 2000). Although the government was making moves toward more regional autonomy, even under Suharto, there was *de facto* autonomy already in the provinces given the shaky hold the new national leaders had on power.

Regional Autonomy and 'Illegal' Logging[11]

'Illegal' logging has been a continual but generally low-capacity activity throughout Indonesia even prior to the current 'reform' era (McCarthy 2000). In the Danau Sentarum National Park area, as elsewhere, the cutters were locals who received capital for logging from legitimate logging companies. These

---

[6] The companies were more constrained in dealing with each other. For example, in the early 1990s, one concessionaire was heavily fined and its local operations shut down for a time after it had been caught cutting trees within another concession.

[7] The highest one-time payment I know of was Rp300 000 (or US$150 at the time) in the early 1990s to a *kepala dusun*.

[8] These included building roads to longhouses (some of which were inconvenient for and little used by locals), helping to repair bridges, providing electric generators, and supplying limited amounts of lumber for school buildings.

[9] The *adat* leaders had divided loyalties on multiple fronts — being approved in office by the government, receiving honoraria from timber companies, and, in the Lanjak area, being concession holders.

[10] The Danau Sentarum conservation project during the mid-1990s was another aspect of this general trend toward increased outside claims on local resources. This has been covered in detail elsewhere (for example, Colfer et al. 1997; Harwell 2000).

[11] I am indebted to Heri Valentius for some of the information presented in this section. My own fieldwork in April–May 2000 was conducted in collaboration with the Centre for International Forestry Research project, 'Local People, Devolution and Adaptive Collaborative Management of Forests', and the following description reflects events at that time. Much has happened since, but cannot be covered here (see Wadley and Eilenberg 2005).

companies would then buy the wood — primarily *meranti* (*Shorea* spp.) — without official documents. Since *krismon*, however, the level of 'illegal' logging has increased in the area and the flow of 'illegally' cut timber has shifted dramatically, some still going to licensed saw and pulp mills, and some going directly into foreign markets.[12]

Over the last several years in the upper Kapuas borderland, local communities and Malaysian financiers have been the chief players, rather than the Indonesian concessionaires. In February 2000, there were no fewer than 12 small financiers (known in Indonesia as *cukong*, and locally as *tauke* or *tukei*) from Sarawak operating in locations along the border from Nanga Badau to Lanjak.[13] Their numbers continued to grow. Six of these *tukei* built substantial sawmills near the main government road that runs to the north of the national park, and the area being logged expanded to accommodate these sawmills.

It appears that the economic crisis has provided an excuse to allow these activities to continue. Numerous local people have said that communities and the *bupati* (district heads) agreed to let Malaysian logging companies into the area because people were not able to make a living because of *krismon*. However, even outside of logging, the economic crisis has probably had a mixed impact on their livelihoods (Sunderlin et al. 2000).

For one thing, since early 1997, locals benefited from a rise in pepper prices (Figure 6.2). This was particularly the case for those who already had well-established pepper gardens prior to the boom.[14] Although farmers close to the border have relied on selling pepper into Sarawak, even prior to the boom, recently improved connections across the border (Wadley 1998) have allowed smallholders further afield to take their produce into Malaysia for sale. In fact, I was told that if they were to sell pepper to (more distant) regional markets such as Sintang, the traders would eventually sell it into Malaysia anyway. However, the 2000 boom had reached its end as production increased throughout Indonesia and elsewhere, driving prices down.[15]

---

[12] Currey and Ruwindrijarto 1999, 2000; Gellert 2000; Cohen and Lembang 2000; *Riau Pos,* 23 June 2000.

[13] McCarthy (2000: 5–6) defines *cukong* as the entrepreneur with capital and *tauke* as the *cukong's* agent who carries out the logging operation. In the Iban areas, *tukei* are, by and large, the entrepreneurs (Richards 1988: 398).

[14] In fact, one man from the Badau area (where pepper smallholdings are long-established) jokingly asked me to use my 'considerable influence' on the world economy to reduce the value of the Rupiah even more as he had profited handsomely at the height of *krismon* by selling his pepper across the border.

[15] *Pepper Market Review,* July–September 2000; *International Pepper Community,* 28 April 2000.

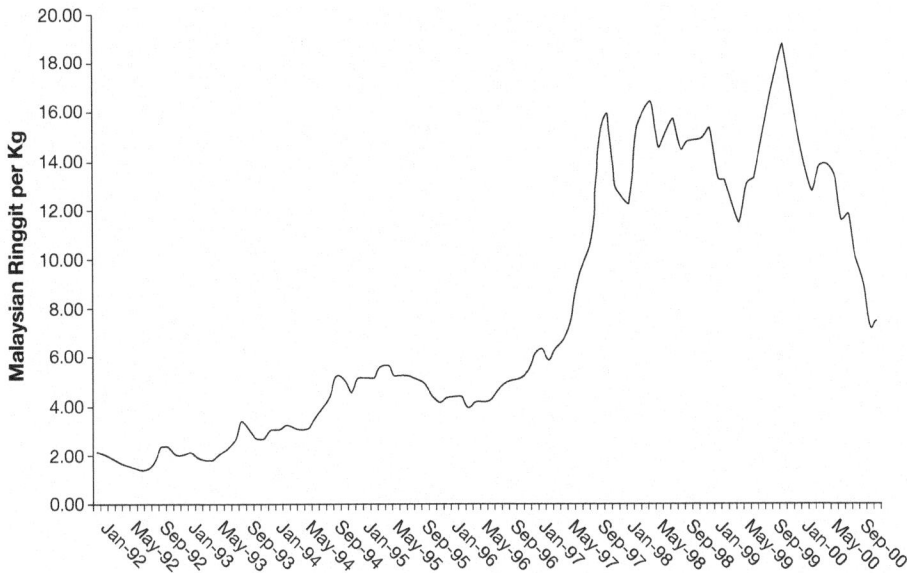

Figure 6.2. Black pepper prices in Kuching, Sarawak (1992–2000)

Source: http://www4.jaring.my/sarawakpepper/stat7.htm

In addition, there has been continued labour migration to jobs in Sarawak, and if pepper prices continue to decline, many local men may return to this option (Wadley 1997, 2002). There have also developed local vegetable markets where women from nearby communities sell produce on any day of the week to feed the expanding, non-farming populations of Lanjak and Badau. Prostitutes from outside the area catered to the truckers and loggers, and new shops, cafes, bars and *losmen* have proliferated in the market towns.

Community Cooperatives

Much of the logging carried out in 2000 was through community cooperatives.[16] Many communities along the northern periphery of the national park belong to a cooperative with several others, often outside *desa* (administrative village) boundaries, while a few communities are independent, such as those dealing with the oil palm company in the Badau district. The alleged aim of these cooperatives has been to facilitate joint development projects among the member communities, but so far the only activity that cooperatives have engaged in has been logging. However, in the future, when all the marketable timber is gone, they may shift to plantation crops.

Cooperatives appear to vary in organisation between two general types: 'directly liaised' and 'indirectly liaised' with timber companies and/or their

---

[16] They operated under Article 10 of the now-superseded Regulation No. 6/99 on *Forest Utilisation and Forest Product Harvesting in Production Forests* (see Casson, this volume).

contractors.[17] The 'directly liaised' cooperatives have included several communities organised by educated residents, often the *kepala desa* (administrative village head) and *kepala dusun* (administrative sub-village head) or other resident leaders. These people negotiate and work with timber company representatives and their contractors when the logging begins. They receive modest commissions and honoraria (the latter being something carried over from the earlier concession system). They work with close relatives in the government to ease the process of obtaining certain permits, and these civil servants also receive modest commissions.[18] These cooperative leaders have said that, under the new system, they can shop around and look for the best deal for their communities.

The 'indirectly liaised' cooperatives have been organised by local educated élites. These liaisons are not resident in the cooperative communities, even though they have kin ties. Some of them have prior ties to the timber industry and all have good connections in local, regional and provincial government; some even have close, pre-existing kin and business ties to timber company bosses in Malaysia. These liaisons have served as gatekeepers, even more so than their counterparts in the 'directly liaised' cooperatives. They control cooperative members' access to information and to the timber company representatives and contractors. The process has thus been less open, and their fees and commissions much higher.[19] Some have also received salaries from the timber companies — in one case to the value of Rp1.5 million (US$180).

Bicycle Logging

The forests being logged have been mainly along the northern periphery of the national park, and in some cases into the northern buffer zone. Timber cutting has also occurred within the park. Most of the logging crews come from Sambas on two-month contracts; the workers are Malay and their overseers are Sambas Chinese who have their own business and family connections with the Malaysian *tukei*. For the most part, locals have not been hired, although in one case local Iban women were being trained as scalers, while some men have worked as truck drivers or tractor operators. One of the main reasons for the lack of locals in the work force is not because the *tukei* have refused to hire them, but rather that the locals were used (even before *krismon*) to getting much higher pay in Malaysia, often working at the same jobs. Sambas crews have been cheaper for the *tukei*.

---

[17] These are my own terms; locals do not appear to make any terminological distinction between the two.

[18] In one case, two men received Rp70 000 (US$8.30) each for helping their home community obtain its cooperative permit.

[19] One such liaison received Rp500 000 (US$60) for processing cooperative permits.

The main method of logging in 2000 may be an excellent example of low-impact timber harvesting, with the heaviest machinery being chainsaws and bicycles. Sawyers fell selected trees and cut them into *balok*.[20] These beams are then loaded onto bicycles that are heavily reinforced, each having two front forks made and welded by the bike owners themselves. The bicycles, carrying one or two beams on either side, are pushed along a track from the cutting site to the main road (in the north of the park) or waterway (in the eastern part). The track is made of a small-diameter roundwood frame about a metre and a half wide, on which are laid two tracks of end-to-end planks (each 10–13 cm wide). The worker walks along one track while pushing the laden bicycle along the other. After the beams have been unloaded, the worker pedals the bicycle back to the cutting site, skilfully balancing along a single track without putting his feet down.[21] The *balok* are transported by truck along the government road and across the border into Sarawak along the very same route used for centuries. (This was before sawmills along the road were fully operational — now rough-milled lumber is being carried instead of *balok*.)

Commissions and Fees

Cooperatives receive commissions for the wood cut in their forests. The commission promised to one cooperative was Rp25 000 per ton (US$2.52 per cubic metre), while another was given 20 Malaysian ringgit per ton (US$4.46 per cubic metre).[22] The difference here was due to the distance to the border, and thus higher transport costs from the first cooperative. In the first case they were paid Rp1.5 million (US$180) for about 70 cubic metres. Four men who worked as community representatives with the outside liaison received Rp850 000 (US$100) to divide among themselves. The principal community representative was given an under-the-table fee for his continued good service — a practice reminiscent of the honoraria given to community leaders under the old concession system.

These fees and commissions were certainly much more than communities had ever received from logging companies in the past — locals are still bitter about their lack of profit from past logging. Once taken across the border, however, the *balok* are milled and exported to Japan, Taiwan and Hong Kong, and the

---

[20] The beams were of various sizes ranging between 10 x 6 x 420 cm and 24 x 24 x 420 cm. The tree species were *meranti* (*Shorea spp.*), *ramin* (*Gonystylus* spp.), *kelansau* (*Dryobalanops oblongifolia*), *bedaru* (*Cantleya corniculata*), and *mabang* (*Shorea* sp.). Not infrequently *tengkawang* (*Shorea macrophylla*) was cut as well, even though it is a protected species, both by national law and local *adat*.

[21] The bicycles have no brakes other than a thick strip of rubber from an automobile tyre mounted between the frame and the back tyre; this is pushed with the sole of the feet against the back tyre to slow the bicycle down.

[22] One British ton = 1.189 cubic metres. The local reliance on measurements in British tons reflects the dominance of Malaysian *tukei* as well as the fact that border Iban are more comfortable with it, many of them having worked for decades in the timber industries of Sarawak and Sabah.

cooperative's commissions amount to about one per cent of the prices realised by Malaysian lumber exporters (US$340 per cubic metre on average).[23]

Questions About the Cooperatives

There are some curious features about these community cooperatives: all appeared to have proper permits issued through the regional office for development cooperatives, but there was no indication that logging profits were divided among the member communities — rather, each community appeared to be acting independently and receiving independent commissions. Additionally, in at least one case of heated disputes between communities over forest land, the disputing communities belonged to the same cooperative! This evidence suggests that timber companies and their local liaisons have been using the cooperative permitting system and 'politically correct' rhetoric about community cooperatives to gain access to forest now in the hands of communities.

In the past, communities located within the concessions had little power over their forests. Since 1998 even Indonesian timber companies with concessions elsewhere in the province have hired community negotiators and public relations officers to deal with local community demands for more compensation. In the national park area, one community negotiated 11 times with the oil palm company (involved in timber cutting) to improve the conditions of their agreement (personal communication, R. Dennis, October 1999). With all but one concession having been terminated, both foreign and domestic timber companies have had to cooperate more publicly with local communities, and cooperatives have been the way in which to do this. Locals see cooperatives as a means to derive some benefit from the forests that have in the past, and might again, become alienated from their control (McCarthy 2000: 9).

Each cooperative has had a permit for establishing itself, and those engaged in logging claimed to have permits from the Ministry of Forestry and Estate Crops, but few permits conformed to existing regulations for legal timber cutting. Those building sawmills (some very small operations, others very large) have claimed to have permits from the Department of Industry. I have only seen cooperative permits, although some cooperatives may indeed have had forestry and industry licences as well. Local businessmen said it was the lack of permits from the Ministry of Trade to export the wood across the border into Malaysia that made the logging 'illegal'. They said it was not illegal logging *per se*, but illegal export, and the reasons for this lay with the *de facto* regional autonomy.

---

[23] At http://www.timber-market.com/tradarea/sample.html prices are listed for various types of Malaysian hardwoods (i.e. $480, 500, 520, 450, 270, 210, 200, and 110 per cubic metre). These provide the average price cited here.

Local Autonomy

The power vacuum created since the fall of Suharto has led to a *de facto* regional autonomy, with regional and local officials being reluctant or unwilling to implement and enforce existing regulations.[24] This has resulted in what local people (and especially local businessmen) have seen as greatly increased corruption (*Indonesian Observer*, 26 November 2000). One businessman claimed that his cooperative lacked a trade permit because he and his *tukei* refused to pay a bribe of Rp15 million (US$1 775) to the Pontianak trade office issuing the permit.

The *tukei* and their liaisons were said to regularly pay off local police, military, *camat* (sub-district heads) and even officials in the *bupati*'s office. In exchange these civil servants turned a blind eye to the logging and daily export of wood across the border and would act surprised whenever a reporter arrived to ask them about the logging. Some local residents unconnected to logging have been increasingly bitter about this corruption. Again they see the wealth of their forests (and in some cases the land itself) going to outsiders, despite increased community involvement in the process. Some have also been angry over what they saw as duplicity on the part of the oil palm company, which promised plantations but was only contracting out for logging.

A local businessman described the situation this way: in the past, under the old concession system, it was your connections to power — to Jakarta — that determined the granting of a concession. Since *krismon*, with government in disarray, it has been the bold and the *berani* who have been favoured, those willing to ignore the rules and pay off local officials. In the past, logging that was unauthorised by the central state would have been shut down quickly. However, there now is evidence to the contrary, with local and foreign businessmen taking advantage of government disorganisation and increased corruption, and in at least some cases, the currently popular cooperatives.

In 2000, local residents and businessmen were looking forward to formal regional autonomy, but they said the cost of doing business would increase with even greater corruption. The potential for severe environmental degradation also appeared to be a consistent worry associated with formal autonomy, as was the potential conflict over its implementation and meaning.[25] Logging has continued at its current accelerated pace, with operations expanding into areas further from the border, and particularly along the main government road. In 2000, some 'indirectly liaised' cooperatives were looking to expand to communities beyond the Embaloh River and into the Kapuas. Once most of the

---

[24] This was certainly a problem in the past (McCarthy 2000), and possibly even more so now.

[25] Buising 2000; Down to Earth 2000; Soedibjo 2000; *Jakarta Post*, 23 December 2000; *Kompas Cyber Media Online*, 8 November 2000; *Akcaya Pontianak Post Online*, 29 November 2000.

marketable timber has been cut (and most agree there is currently enough for five years), conversion to oil palm plantations would likely follow.

Local Disputes

Under the new system, local communities have been more emboldened and empowered to deal independently with economic change. However, the way this has been done has divided some communities (for example, between those who do and do not want to participate in an oil palm scheme), and it has in some cases led to the realignment of local communities' territorial boundaries.[26] There have been a number of community disputes over forests since this logging boom began. In at least one case, the dispute was over forest land that had never been part of any traditional community territory. In some instances, the disputes were settled by a cockfight, with the winning community gaining possession of the disputed land and forest. Locals recognised all this as a rush to make claims on timbered land so that they would receive a portion of the logging profits.[27]

A dispute between the Iban communities of Lanting and Kelayang[28] on the lower Leboyan River began when men from Kelayang confiscated three chainsaws belonging to workers cutting timber on what the loggers took to be Lanting land. Kelayang residents claimed that the timber cutting went beyond Lanting boundaries and into their own land. Residents of Lanting denied these claims, saying the area being cut was well within their community territory. Subsequently, Kelayang residents used red paint to mark trees along what they claimed was their boundary, but Lanting did not agree with this. Further, Kelayang was making use of maps drawn under a conservation project's community-mapping program, although in this instance, Lanting was not consulted about the original map-making program.

An attempt was made to settle the case in a meeting at Kelayang, where the *kepala desa* resides. The confiscated chainsaws were returned after being redeemed with money by the loggers, but no agreement on the boundary could be reached as Lanting felt the boundary set by Kelayang took away too much of Lanting's land. Because this case could not be settled by the communities themselves, it was taken to the next highest level of *adat* adjudication; the Iban *temenggong* of Kecamatan Batang Lupar. If no agreement could be reached, both parties agreed to a traditional cockfight to finally settle the issue.

There has also been a heated dispute to the north of Danau Sentarum National Park between two other Iban communities. In late 1998, Sarawak *tukei* began working with people from one longhouse (without a written agreement) to cut

---

[26] The Danau Sentarum conservation project also gave impetus to some shifting territorial claims through its ambitious community-mapping program (Dennis 1997).

[27] Elsewhere *otonomi daerah* threatens to lead regional governments into conflict over their own boundaries (*Kompas Cyber Media Online*, 19 January 2001).

[28] These are pseudonyms.

forests that the community claimed as its own. The people of the other longhouse claimed otherwise, and this eventually led the two communities to settle the matter by a cockfight in April 2000. The first community lost the match and its claims on the forest. The dispute over this land (which had never really been part of the traditional territory of either community) has created a good deal of bitterness on both sides.

In another case, one community refused to cooperate with logging operations, and the *tukei* deliberately created trouble between this community and another more cooperative community. The *tukei* gave shotgun shells to people from the cooperating longhouse in order to intimidate their neighbour. People from the first longhouse became aware of this situation and fired shotguns at the sawmill camp owned by the *tukei* and located near the cooperating longhouse. Several people from the first community wrote a letter rejecting such foreign intrusions and submitted it to government officials with the stipulation that they would act alone if the government did not deal with the problem.[29]

These disputes have been damaging to relations between often closely related communities. During post-harvest rituals, for example, one community normally invites related longhouses to attend, and this has important social and economic integrative functions (Dove 1988). After all this, however, people from disputing communities will be reluctant to attend each other's rituals, and in the case of Lanting and Kelayang, people from the former community have been delayed in building a new longhouse because the present boundary problem has not yet been settled. Unsettled disputes are said to produce supernaturally hot (*angat*) conditions, and ritually sensitive activities such as house building must be avoided during such times. Close kin relations have been (temporarily?) attenuated over access to valuable timber, with at best mediocre compensation.

International Disputes

This logging activity has further challenged the exclusivity of the Indonesian–Malaysian border,[30] yet locals were of the strong opinion that nothing would be done because local government officials, military and police are paid off by the timber bosses or their representatives. Occasional news reports of 'illegal' logging and smuggling of cut timber have appeared in the national and regional press, but efforts to prevent it have tended to be very meagre.[31] Along

---

[29] The Iban have proven themselves quite capable of taking matters into their own hands. For example, they effectively resisted military efforts in the 1960s–70s to confiscate their shotguns used for hunting. And in December 2000, a group of some 400 Iban men from the Badau area stormed a courthouse in Putussibau and killed a man who was on trial for murdering an Iban money changer earlier that year (*Kompas Cyber Media Online*, 14 December 2000).

[30] This is also an issue along the Sabah-East Kalimantan border (*Suara Pembaruan*, 3 October 2000).

[31] *Bisnis Indonesia*, 12 September 2000; *Jakarta Post*, 28 May 2000; *Akcaya Pontianak Post Online*, 28 November 2000.

with other cross-border activity, this 'illegal' logging has threatened to lead to international disputes between Indonesia and Malaysia.[32]

There have been cooperative Indonesian–Malaysian efforts to survey the border to show whether or not cutting by Malaysians was illegal.[33] In addition, plans to open an official border post at Nanga Badau might help to prevent illegal exports and may encourage taxing of goods going across the border; on the other hand, it might also lead to increased corruption.[34]

Discussion and Conclusion

Recent reform policies in Indonesia have led to the termination of numerous timber concessions and to legislation providing local communities with the opportunity to establish cooperatives for their own development. This has allowed communities to take more control of their forests. In addition, the power vacuum created by the fall of Suharto's New Order regime has left local and regional governments in some confusion, resulting in a simultaneous increase in local empowerment (as the police and military have lost a good deal of legitimacy) and official corruption. At the time, the implementation of formal *otonomi daerah* was still very much in question; however, even before its deadline for implementation in 2001, a *de facto* regional autonomy existed.

In the borderlands of West Kalimantan, these changes have been visible in the heavy involvement of Sarawak timber companies with local community cooperatives in logging forests under the ostensible control of communities.[35] This was nowhere more apparent than in the upper Kapuas borderland inhabited by the Iban, where historical routes of trade are being used to transport wood across the international border. Although officially considered smuggling, the practice of paying local officials, police and military has made this activity clearly visible. It was uncertain how the eventual implementation of *otonomi daerah* and the opening of an official border crossing in the area would affect this activity.

Local communities have seen this time as quite positive, as they have the power to manage their resources for and by themselves. Their involvement with Malaysian *tukei* has caused them little alarm, given their own position as borderlanders. However, they have been worried about future resource competition from timber and oil palm companies that might gain legally binding

---

[32] *Kompas Cyber Media Online*, 4 July 2000; *Suara Pembaruan*, 11 September 2000.

[33] *Kompas Cyber Media Online*, 10 June 2000; *Akcaya Pontianak Post Online*, 16 November 2000. This disputed area, formerly part of Kabupaten Sambas and now in the new Kabupaten Bengkayang, was subject to a special Dutch–English border survey and agreement in the late 1920s (Netherlands Indies 1930).

[34] *Kompas Cyber Media Online*, 3 September 1999; *Kompas Cyber Media Online*, 5 January 2001.

[35] In fact, local communities appear to have been the only real managers of forests in the area (Colfer et al. 1997).

licences to their forests. This threat appeared to be one factor driving local cooperatives to allow logging in their forests. Another factor was the continuing economic crisis, and although there have been local economic alternatives available (such as pepper gardening and labour migration), logging appeared to be a quick way to earn ready cash, particularly as local Iban tended not to be doing the actual work. The power of communities to benefit more from the logging has been checked in part by the position of local élites who served as liaisons between *tukei* and communities. Even those local communities that deal directly with *tukei* may have been hindered in negotiations by their general lack of information about the value of their timber.

Logging practices appeared to be relatively low-impact, involving nothing more than chainsaws and bicycles, and given adequate control, local communities might be able to prevent widespread damage to their forests. This low-impact method might offer a chance for forest management at the local level, while still allowing occasional (and very long-cycle) logging. Under the circumstances of the time, however, this prospect was probably not very good. The demands on local timber may simply have resulted in more bicycle-logging crews in the forest, which is particularly worrisome for the long-term health of the forests. Extraction of high-quality timber has immediate and ancillary effects on surrounding biodiversity, and impoverishment of the forests may well lead to an impoverishment of local people.

Iban have relied heavily on their forests for swidden rice farming and numerous non-timber forest products. One study determined that Iban who are unaffected by timber cutting and related forest destruction purchased only about nine per cent of their foods; the remainder came from fields and forest (Colfer et al. 2000).[36] As with most poor people in Indonesia, the Iban have tended to rely on a mixed strategy for household livelihood. In addition to the all-important rice farming, they have collected forest products and garden vegetables for sale and home consumption; they have cultivated rubber and pepper as cash crops, and they have engaged in circular labour migration. Their position on this borderland adjacent to a more prosperous and politically stable neighbour and their identity as a partitioned ethnic group has meant that part of that mixed strategy lies across the border, where they have not only found temporary employment but also occasionally places to permanently migrate. Logging is part of this general circumstance and part of the mixed strategy.

The borderland character and the current logging activity in this area was reflected clearly in something I encountered along the government road between Nanga Badau and Lanjak. There, where a bicycle-logging track ended and *balok* were piled for pick-up, a local had painted a sign reading 'CV Munggu Keringit

---

[36] In contrast, neighbouring Malay communities dependent on fishing, and without the same access to forest resources, purchased 59 per cent of their foods.

Sdn Bhd'. This very effectively summed up the ambiguous position of borderland residents engaged in cooperative logging: 'CV' stands for 'limited partnership' in Indonesia (from *Commanditaire Vennootschap* in Dutch), while 'Sdn Bhd' stands for virtually the same thing in Malaysia (from *Sendirian Berhad* in Malaysian). As such a designation had no legal standing, it was obviously intended as a joke. But the sign conveyed very well the message that these borderlanders would continue to look to both sides in their efforts to secure a livelihood. Their position and identity as borderlanders must be given consideration in any search for income-generation alternatives that are economically and environmentally viable (Wadley and Eilenberg 2005).

The strengthening of local *adat* is often touted as an important means of empowering local peoples to deal with outside pressures on their resources and much official lip service is paid to it (see Eghenter, this volume). However, its effectiveness may often be overestimated. While providing some legitimacy to *adat* may heighten local self-esteem, *adat* by itself may be incapable of dealing with the issues it faces, particularly where third-party support is non-existent. A general consensus is needed among communities for devising and implementing effectively binding rules and sanctions, but *adat* leaders as well as the people they represent tend to have many divided loyalties themselves. *Adat* should not be expected to function adequately under these conditions (McCarthy 2000).

Local NGOs can give advice on, and provide critical services in, several areas: Indonesian natural resource law; regulations on international investment and relations; ways to register community land; and negotiation tactics and strategies. However, given the extremely weak judiciary and law enforcement, knowledge of laws and statutes may not provide real power in the courts, but rather may become useful in negotiations with companies. This is probably more viable than supporting local *adat*, especially if the NGOs involved are formed by people from the communities involved, with their own families' interests at stake (Clarke et al. 1993). Some caution is warranted in a blanket embrace of NGOs as some may actually 'take advantage' of international support for local organisations.

The payment of compensation for not logging would be a costly exercise and would require substantial outside funding, but paying local communities to protect their forests may be an important option (personal communication, E. Harwell, March 2000). Such a program, however, would have to be long-term, with adequate monitoring to determine continued compliance by the community. It might be done in conjunction with management of conservation areas such as national parks (Hamilton et al. 2000).

Regulations such as No. 6/99 on *Forest Utilisation and Forest Product Harvesting in Production Forests*, which allows for the issue of Forest Product Harvesting Rights over a maximum of 100 hectares a year, certainly did very little to promote long-term preservation of forest resources and actually encouraged local communities to cut their forests quickly for immediate profit. Although this regulation has been suspended (see Casson, this volume), there is no indication that logging has subsequently slowed. Indonesian national and local NGOs have played an important role in lobbying for a change to these laws.

References

Alvarez, R.R., 1995. 'The Mexican–US Border: The Making of an Anthropology of Borderlands.' *Annual Reviews in Anthropology* 24: 447–470.

Anonymous, 1856. 'Journal of a Tour on the Kapuas.' *The Journal of the Indian Archipelago and Eastern Asia* 1: 84–126.

Asiwaju, A.I., 1976. *Western Yorubaland under European Rule, 1889–1945.* Atlantic Heights: Humanities Press.

———, 1983. *Borderlands Research: A Comparative Perspective.* El Paso: University of Texas (Border Perspectives Paper 6).

——— (ed.), 1985. *Partitioned Africans: Ethnic Relations Across Africa's International Boundaries, 1884–1984.* London: C. Hurst.

Boggs, S.W., 1940. *International Boundaries: A Study of Boundary Functions and Problems.* New York: Columbia University Press.

Buising, T., 2000. 'A Century of Decentralisation.' *Inside Indonesia* 63 (Jul–Sep 2000). Viewed 23 February 2006 at http://www.insideindonesia.org/edit63/buising4rev.htm

Clarke, H.R., W.J. Reed and R.M. Shrestha, 1993. 'Optimal Enforcement of Property Rights on Developing Country Forest Subject to Illegal Logging.' *Resource and Energy Economics* 15: 271–294.

Cohen, M. and P. Lembang, 2000. 'Wood Cuts: Illegal Logging Could Stem the Flow of Aid to Indonesia.' *Far Eastern Economic Review* 163(4): 20–22.

Colfer, C.J.P. and R.L. Wadley, 1996. 'Assessing 'Participation' in Forest Management: Workable Methods and Unworkable Assumptions.' Bogor: Centre for International Forestry Research (Working Paper 12).

———, E. Harwell and R. Prabhu, 1997. 'Assessing Inter-generational Access to Resources: Developing Criteria and Indicators.' Bogor: Centre for International Forestry Research (Working Paper 18).

———, R.L. Wadley, A. Salim and R.G. Dudley, 2000. 'Understanding Patterns of Resource use and Consumption: A Prelude to Co-management.' *Borneo Research Bulletin* 31: 29–88.

Currey, D. and A. Ruwindrijarto, 1999. 'The Final Cut: Illegal Logging in Indonesia's Orangutan Parks.' Viewed 23 February 2006 at www.telapak.org/publikasi/download/The_Final_Cut.pdf

———, 2000. 'Illegal Logging in Tanjung Puting National Park: An Update on the Final Cut Report.' Viewed 23 February 2006 at www.salvonet.com/eia/old-reports/Forests/Reports/tanjung

Dennis, R.A., 1997. 'Where to Draw the Line: Community-Level Mapping in and Around the Danau Sentarum Wildlife Reserve.' Bogor: Wetlands International, PHPA, ODA.

Dove, M.R., 1988. 'The Ecology of Intoxification among the Kantu' of West Kalimantan.' In M.R. Dove (ed.), *The Real and Imagined Role of Culture in Development: Case Studies from Indonesia.* Honolulu: University of Hawaii Press.

Down to Earth, 2000. 'Regional Autonomy, Communities and Natural Resources.' Viewed 23 February 2006 at www.dte.gn.apc.org/46RAC.htm

Gellert, P.K., 2000. 'Dynamics of Indonesia's Timber Industry after the Financial Crisis and after Suharto: Contradictory Pressures on Loggers, Forests and Local People, Especially in Kalimantan (Borneo).' Paper presented at the sixth biennial conference of the Borneo Research Council, Kuching, Sarawak, 10–14 July.

Hamilton, A., A. Cunningham, D. Byarugaba and F. Kayanja, 2000. 'Conservation in a Region of Political Instability: Bwindi Impenetrable Forest, Uganda.' *Conservation Biology* 14: 1722–1725.

Harwell, E.E., 2000. The Un-Natural History of Culture: Ethnicity, Tradition and Territorial Conflicts in West Kalimantan, Indonesia, 1800–1997. New Haven (CT): Yale University (Ph.D. thesis).

Kater, C., 1883. 'Iets over de Batang Loepar Dajakhs in de "Westerafdeeling van Borneo" [A Point about the Batang Loepar Dayaks in the "Western Division of Borneo"].' *Indische Gids* 5: 1–14.

Kielstra, E.B., 1890. 'Bijdragen tot de Geschiedenis van Borneo's Westerafdeeling [Contribution to the History of Borneo's Western Division].' *Indische Gids* 12: 1090–1112, 1482–1501.

Mackie, J.A.C., 1974. *Konfrontasi: The Indonesian–Malaysian Dispute, 1963–1966.* Kuala Lumpur: Oxford University Press.

Martinez, O.J., 1994. *Border People: Life and Society in the U.S.–Mexico Borderlands.* Tucson: University of Arizona Press.

McCarthy, J.F., 2000. '"Wild Logging": The Rise and Fall of Logging Networks and Biodiversity Conservation Projects on Sumatra's Rainforest Frontier.' Bogor: Centre for International Forestry Research (Occasional Paper 31).

McKeown, F.A., 1983. The Merakai Iban: An Ethnographic Account with Especial Reference to Dispute Settlement. Melbourne: Monash University (Ph.D. thesis).

Netherlands Indies, 1930. 'Verdrag tot Nadere Vaststelling van een Gedeelte der Grens tusschen het Nederlandsch Gebied op het Eiland Borneo en Serawak [Treaty until Further Settlement of a Section of the Border between the Dutch Region of the Island of Borneo and Sarawak].' Batavia: Landsdrukkerij (Staatsblad van Nederlandsch-Indië 375 [Statute Book of the Netherlands Indies 375]).

Niclou, H.A.A., 1887. 'Batang-Loepars — Verdelgings-Oorlog: Europeesch-Dajaksche Sneltocht [The Batang Loepars' War of Extermination. European-Dayak Rapid Expedition].' *Tijdschrift voor Nederlandsch Indië* 1: 29–67.

Pfeiffer, I., 1856. *A Lady's Second Journey Round the World*. New York: Harper and Brothers.

Pringle, R., 1970. *Rajahs and Rebels: The Ibans of Sarawak under Brooke Rule, 1841–1941*. London: Macmillan.

Richards, A., 1988. *An Iban-English Dictionary*. Petaling Jaya: Oxford University Press.

Rumley, D. and J.V. Minghi (eds), 1991. *The Geography of Border Landscapes*. London: Routledge.

Soedibjo, B.S., 2000. 'Otonomi Daerah dan Konflik Lingkungan [Regional Autonomy and Environmental Conflict].' *Media Indonesia*, 26 September.

Soemadi, 1974. *Peranan Kalimantan Barat dalam Menghadapi Subversi Komunis Asia Tenggara [The Role of West Kalimantan Facing the Communist Subversion of Southeast Asia]*. Jakarta: Yayasan Tanjungpura.

Sunderlin, W.D., I.A.P. Resosudarmo, E. Rianto and A. Angelsen, 2000. 'The Effect of Indonesia's Economic Crisis on Small Farmers and Natural Forest Cover in the Outer Islands.' Bogor: Centre for International Forestry Research (Occasional Paper 28E).

Wadley, R.L., 1997. Circular Labor Migration and Subsistence Agriculture: A Case of the Iban in West Kalimantan, Indonesia. Tempe (AZ): Arizona State University (Ph.D. thesis).

————, 1998. 'The Road to Change in the Kapuas Hulu Borderlands: Jalan Lintas Utara.' *Borneo Research Bulletin* 29: 71–94.

————, 2000. 'Warfare, Pacification, and Environment: Population Dynamics in the West Borneo Borderlands (1823–1934).' *Moussons* 1: 41–66.

————, 2001. 'Trouble on the Frontier: Dutch-Brooke Relations and Iban Rebellion in the West Borneo Borderlands (1841–1886).' *Modern Asian Studies* 35: 623–644.

————, 2002. 'Coping with Crisis — Smoke, Drought, Flood, and Currency: Iban Households in West Kalimantan.' *Culture & Agriculture* 24: 26–33.

————, 2003. 'Lines in the Forest: Internal Territorialization and Local Accommodation in West Kalimantan, Indonesia (1865–1979).' *South East Asia Research* 11: 91–112.

————, R.A. Dennis, E. Meijaard, A. Erman, H. Valentinus and W. Giesen, 2000. 'After the Conservation Project: Danau Sentarum National Park and its Vicinity — Conditions and Prospects.' *Borneo Research Bulletin* 31: 385–401.

———— and M. Eilenberg, 2005. 'Autonomy, Identity and "Illegal Logging" in the Borderland of West Kalimantan, Indonesia.' *Asia Pacific Journal of Anthropology* 6(1): 19–34.

Chapter Seven

Seeking Spaces for Biodiversity by Improving Tenure Security for Local Communities in Sabah[1]

Justine Vaz

Introduction

With the steady degradation of the world's tropical forests and reduced confidence in the protected-area model, some attention has turned to the potential role that community-claimed forests could play in biodiversity conservation. In Sabah and elsewhere in Asia, the customary lands of upland communities — often comprising tapestries of homesteads and farms, fallowed fields, mature secondary forest and the hinterland of riverine and primary forests — could potentially serve as refuges for threatened biodiversity. With long histories of residence, active use of the forest landscape, and an apparent affinity to the forest, many local or indigenous community lifestyles have been seen to represent a more gentle and peaceable future for tropical forests. Indeed, in recent years various groups have captured international attention by their efforts to defend forest areas that have increasingly come under threat from logging and forest conversion (Hong 1987; Poffenberger and McGean 1993; Baviskar 1995; Colchester 1997) and impressive feats of collective action to restore degraded forest (Poffenberger and McGean 1996; Stevens 1997).

Often the lack of *de jure* rights of ownership to forest areas has proven to be the major stumbling block to these movements. Customary claims are frequently not adequately recognised by modern government administrations, or the same forest resources are classified under the eminent domain of the state (Brookfield et al. 1995: 128). In such instances, strengthening local tenure in collaboration with local residents has been viewed by conservation organisations as a valid endeavour to stem imminent threats to important natural areas. The move to

---

[1] I first became acquainted with the local community in Sabah's Upper Padas region in 1997 during a government consultancy to identify new protected areas. This research was subsequently conducted between January 1999 and March 2001 in the course of establishing the Ulu Padas Community-Based Conservation and Development Project, a joint initiative of WWF-Malaysia and the Ministry of Tourism and Environmental Development. The Ulu Padas experience is now being evaluated away from the field for a Ph.D. in Geographical and Environmental Studies at the University of Adelaide. Consequently, opinions expressed here are my own and may not necessarily coincide with those of WWF-Malaysia. This chapter has benefited from insights into the dynamics of community life and tenure issues provided by Alison Hoare, who conducted an independent investigation of Lundayeh land and forest resource use between September 1999 and October 2000.

lend resources and expertise to such initiatives is also underscored by the belief that this could contribute to the restoration of communal management systems and, in the process, establish spaces where biodiversity and community interests might coexist.

In March 1999, I led a project for WWF-Malaysia to advocate for the conservation of montane forest in the biologically significant Ulu Padas headwaters in Sabah's southwest. The project involved working closely with the Lundayeh community of this area. For years the uncertain status of land ownership had proven to be a significant factor in forest degradation. The intervention in the Upper Padas was intended to tap the potential for securing spaces for biodiversity by seeking greater security of tenure for local residents.

By drawing upon this experiment in building community–conservation partnerships, this case study acknowledges the potential for synergy between strengthening communal tenure and conserving biodiversity. However, field experience shows that it is necessary to modify expectations of the local community's commitment to conservation. Though local people profess a strong affection for and appreciation of the surrounding environment, this alone does not provide sufficient assurance of actions that prioritise conservation. This has probably never been more apparent than in this period of rapid social change where greater access to urban society, systems and mores has had a tremendous influence on highly mutable local aspirations. Greater care is needed in negotiating community–conservation partnerships if outcomes are to have any relevance to both environmental conservation and local people's aspiration for economic development. This chapter discusses some of the ways in which community–conservation partnerships might be based on more explicit arrangements that satisfy the specific interests of the parties involved. It also highlights the value of policy reform and collaborative efforts involving NGOs, communities and government agencies in promoting a land-tenure resolution process that safeguards the long-term wellbeing of both local communities and the environment.

One Landscape, Three Competing Interests

Biogeographic and Conservation Significance

Ulu Padas refers to the headwaters of the Padas River, an area of approximately 80 000 hectares at the southwestern-most tip of Sabah, Malaysian Borneo. This steep mountainous area, with elevations ranging from 915 to 2070 metres, remains among the few parts of Sabah's forest estate with extensive old-growth forest

(Figure 7.1) (Mannan and Awang 1997: 2).[2] Globally, Ulu Padas is of considerable conservation significance. It is believed to rival Mount Kinabalu in terms of plant endemism and species diversity, particularly within pockets of rare *kerangas* or heath forest throughout the area. This area is part of the larger Central Bornean Montane forests, a transboundary ecoregion that extends over the Kelabit Highlands of Sarawak and Indonesia's Kayan Mentarang National Park (Figure 7.2). The contiguous oak-chestnut forest is also believed to support the seasonal migration of the bearded pig (*Sus barbatus*), a major source of meat for Borneo's upland communities (Hazebroek and Kashim 2000).

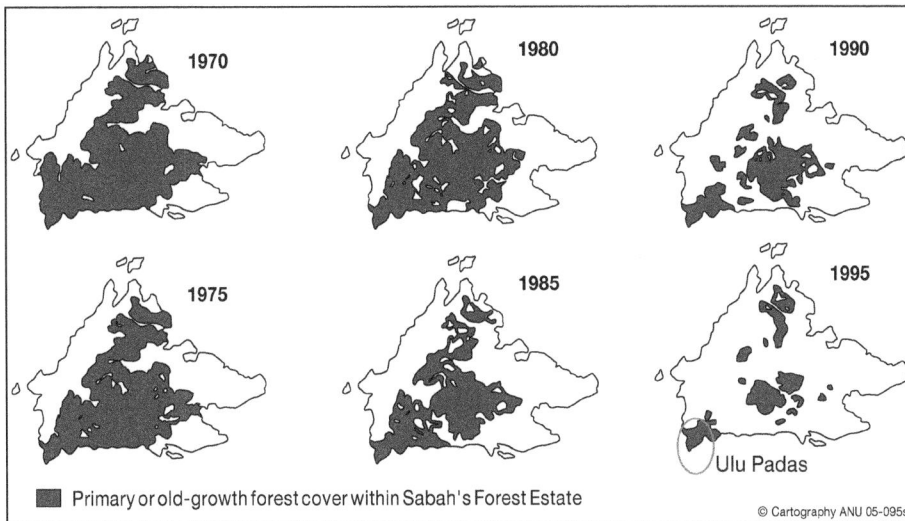

Figure 7.1. Contraction of primary old-growth forest in Sabah's Permanent Forest Estate, 1970–95

Source: Sabah Forestry Department 1997

As early as 1992, Ulu Padas was identified in the Sabah Conservation Strategy as a distinct biogeographic zone warranting inclusion in the state's protected area network. To pursue these recommendations further, in 1997 WWF-Malaysia in association with the Ministry of Tourism and Environmental Development, and supported by the Danish Agency for Cooperation and Development, commenced the 'Identification of Potential Protected Areas' component of the Sabah Biodiversity Conservation Project. In Ulu Padas, botanical collections confirmed early suspicions of biological significance, identifying 11 distinct

[2] In 1997, the then Acting Director of the Sabah Forestry Department reported that between 1975 and 1995, the overall coverage of primary forest in Sabah's forest estate 'dwindled from 2.8 million hectares to about 0.3 million hectares'. In the Commercial Forest Reserves intended for sustainable forest management, old-growth cover was estimated at only 15 per cent in 1996, compared to 98 per cent in 1970.

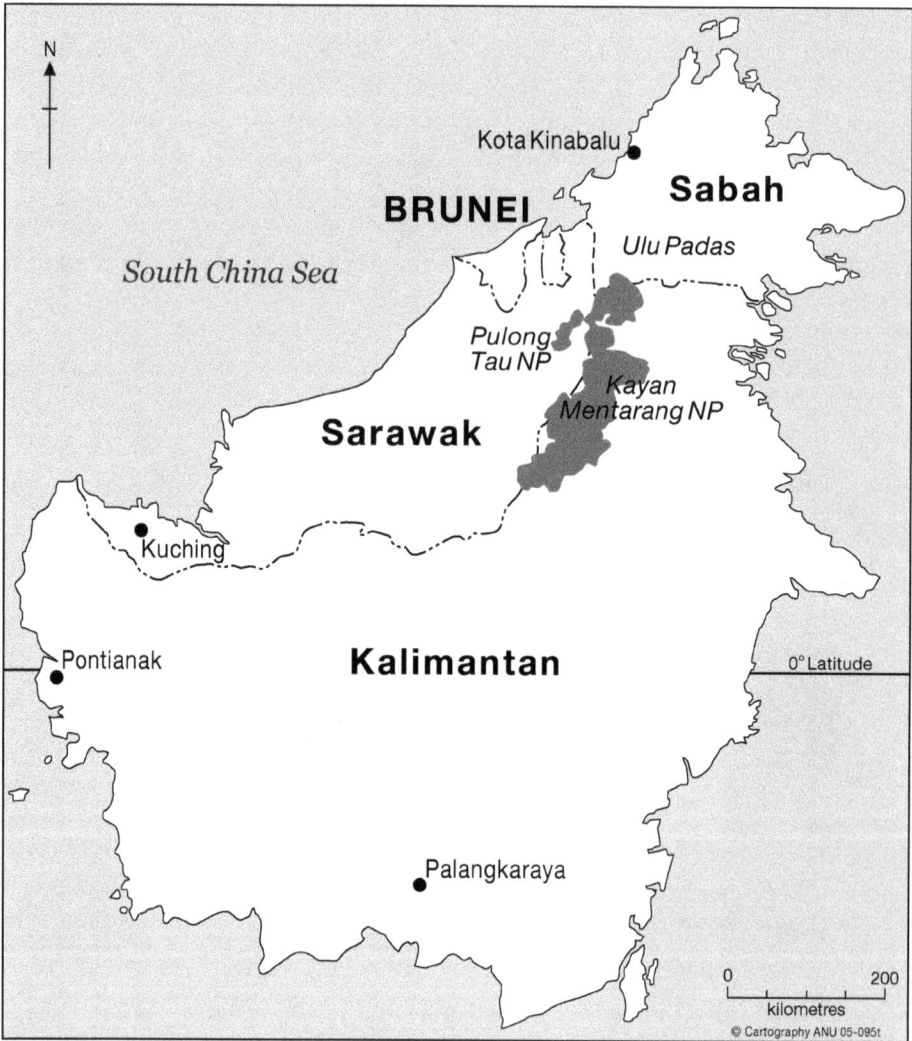

Figure 7.2. Location of Ulu Padas within the Central Bornean Montane Forest ecoregion

forest types and an impressive array of endemic species (Phillipps and Lamb 1997). The combination of high annual rainfall, high elevation and steep terrain was highlighted in recommendations for catchment management. The Padas River supplies water and generates hydroelectric power for the urban and agricultural areas of Sabah's southwestern region (Sinun and Suhaimi 1997; Paramanathan 1998). Social assessments revealed interest and support at the community level for conservation and associated development opportunities (Towell 1997). Community apprehensions that logging in the surrounding Forest Reserves would threaten their way of life also featured frequently in formal and informal discussions. At the conclusion of the 1997 study, stakeholder workshops and discussions were held in order to share this information and seek a common vision for this area, which included recommendations to convert the Ulu Padas Commercial Forest Reserve into a Protected Area (Payne and Vaz 1998).

Community Claims to Land and Forest

The Ulu Padas community comprises two villages with a combined population of approximately 500 people, centred at the mouths of the Pasia and Mio rivers, both tributaries of the Padas (Figure 7.3). The Lundayeh people of Long Pasia and Long Mio are mainly subsistence swidden and wet-rice farmers, although tobacco, coffee, vegetables and fruits are increasingly being planted. Wild game is the primary source of protein and hunting is an integral part of Lundayeh identity. Rivers supply fresh water and fish, and the surrounding forest is an important source of food, medicines, firewood and building materials (bamboo, rattan and wood) (Hoare 2002: 41–73). Local people regularly access forest resources far beyond existing farms and homesteads, particularly for medicines and rare plants that only occur in the pristine forest areas (Kulip et al. 2000). These are also the best hunting grounds. The remoteness of the villages (123 km by logging road from Sipitang) and seasonality of cash incomes make the forest both a lifeline and a safety net for local people.

Generally, the Lundayeh of Long Pasia and Long Mio assert customary claims to land that their forefathers cleared and farmed before them according to the traditional system. They view the area to be their ancestral heartland and see maintaining aspects of their unique way of life as essential to maintaining their ethnic identity. Through their activities, local people reaffirm their long history and connection with the area. Over generations, their agricultural cycles have shaped the environment, developing a mosaic of forest in different stages of regeneration[3] and altering the species composition of amenity forest (Hoare 2002: 152–6). This is also a cultural landscape dotted by burial sites, headhunting monuments, historic foot-trails to neighbouring villages in Sarawak and Kalimantan, and trees and farms planted by ancestors. A rich local folklore explains the formation of rock monuments and striking geological features (Vaz 1999) (see Figure 7.4).

---

[3] In this way, swidden farming is akin to rotational agroforestry and encompasses the management of swidden fields and fallows in multiple stages of development (Peluso 1995: 393).

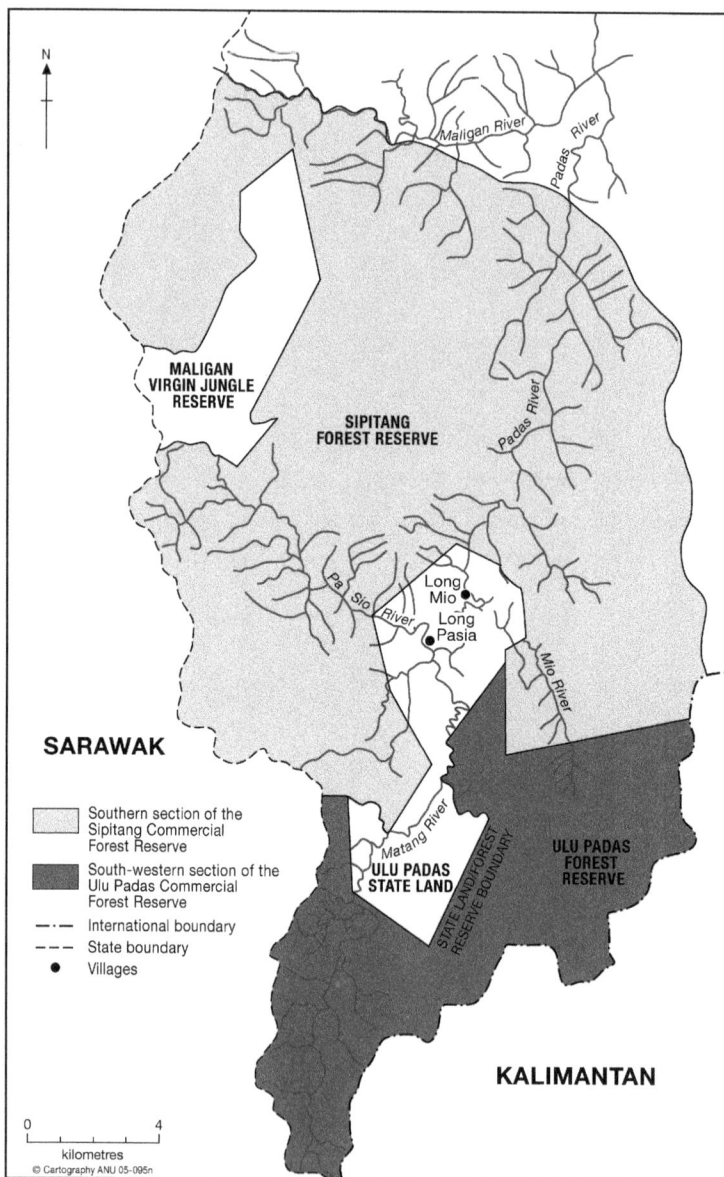

Figure 7.3. Villages and land use classification in the Ulu Padas region

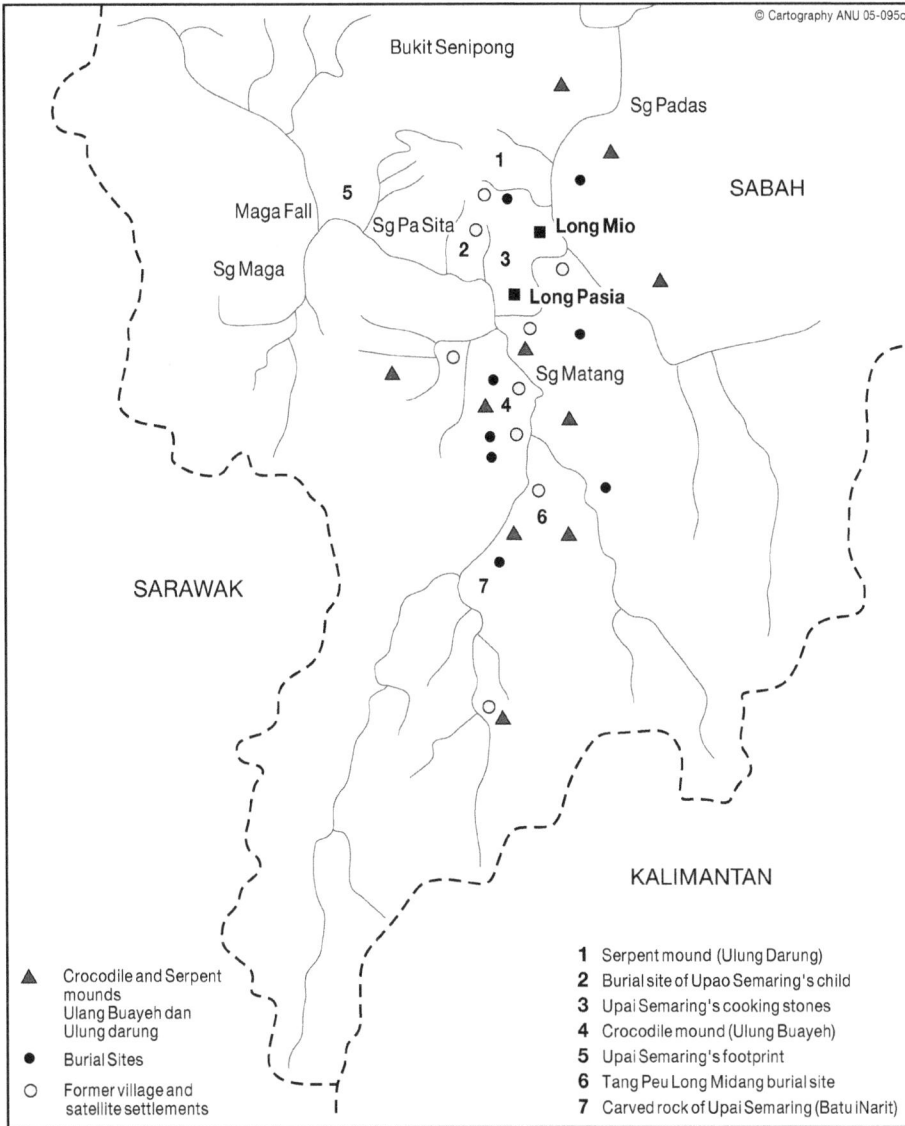

© Cartography ANU 05-0950

Legend:

▲ Crocodile and Serpent mounds
Ulang Buayeh dan
Ulung darung

● Burial Sites

○ Former village and
satellite settlements

1 Serpent mound (Ulung Darung)
2 Burial site of Upao Semaring's child
3 Upai Semaring's cooking stones
4 Crocodile mound (Ulung Buayeh)
5 Upai Semaring's footprint
6 Tang Peu Long Midang burial site
7 Carved rock of Upai Semaring (Batu iNarit)

Figure 7.4. Cultural heritage sites in Ulu Padas State Land

However, in the eyes of the government, despite local people's perceptions of customary claims, only an 'island' of State Land of approximately 12 300 hectares has been set aside for local people to make formal applications for Native Title. The remainder of the Ulu Padas area is classified as Commercial Forest Reserve and is within a Forest Management Unit of close to 290 000 hectares which has been concessioned to Sabah Forest Industries.

Logging Interests in the Upper Padas

Sabah Forest Industries (SFI) is a former state-owned entity managed by majority equity holder Lion Group Holdings since early 1994 (*Asian Timber,* February 2000). In addition, the Ulu Padas Forest Reserve, an area of almost 30 000 hectares proposed as a new Protected Area in the Sabah Conservation Strategy, has been incorporated into a binding 99-year lease agreement (1996–2094) with SFI (see Figure 7.3). SFI's concession is divided into two categories: (1) Industrial Tree Plantation areas, where natural forest is cleared for pulp and paper and replaced with fast-growing species; and (2) areas under Natural Forest Management ,which are meant to be managed for the sustainable harvest of timber according to the state's Forestry Guidelines. SFI's integrated timber complex is the major industry in the nearby town of Sipitang, employing over 2000 people and linked with numerous other contractors and businesses.

Conflict over Forest in Ulu Padas

The imperative to secure community ownership of forests in the Ulu Padas intensified with the profound changes in the surrounding landscape between 1998 and early 1999. By this time, tropical timber was progressively being sourced further in the uplands, more than 100 km from Sipitang. As replanting with Acacia and Eucalyptus had not kept pace with the demand from the mill, logging roads were becoming more and more invasive, penetrating deeper into the forested interior of the Upper Padas. At higher elevations, logging operations targeted the giant Agathis trees of the old-growth montane forest. SFI had become one of Japan's main suppliers of sawn Agathis timber (*Asian Timber,* February 2000). Logging activities within surrounding catchments silted up the tributaries that run through the two valley settlements. Long Pasia's famous 'red river',[4] usually coloured a clear red by the tannins leached from leaf-littered cloud forest, had become the colour of milky tea. Flooding and declining forest resources were also experienced. Long Mio had already been contending with similar problems with the Mio River as a result of logging activity upstream around Muruk Mio, a distinctive peak in the region.

The impacts experienced by the community stimulated a period of heightened environmental awareness and protest, not only about the commercial logging activities in the Forest Reserves (*Daily Express,* 18 April 1999; *The Star,* 19 April 1999; *New Straits Times,* 21 April 1999), but also about the lack of security given to the community's customary lands. The local community argued that their way of life and livelihoods were at risk. 'Allocate an area for the Lundayeh' was the appeal from the President of the Lundayeh Cultural Association of Sabah (*Daily Express,* 11 April 1999). Similar views were expressed at a village meeting, the minutes of which were sent to the Chief Minister's Department.

---

[4] Long Pasia means 'mouth of the red river' in the Lundayeh language.

Our forefathers did not bequeath us wealth of gold and money. Our only inheritance is the land along the banks of the Lelawid and Melabid rivers which they cleared and farmed — this land has been handed from generation to generation. For this reason, we are appealing to the Natural Resources Office for this land to be removed from the Sabah Forest Industries area for us. This land will be divided among the relevant families and also given to village members who do not have land (Minutes from Long Pasia village meeting, 15 September 1998).

At the time of initial WWF-Malaysia dialogues with the community in 1997, the significant reduction in the extent of traditional resource areas and the onset of logging activities were seen to impose unprecedented threats to their environment, economic activities and quality of life. The community had made several attempts to raise their concerns with higher authorities but had little success at obtaining assurances that their customary claim to land and forest in Ulu Padas would be recognised or that logging would be controlled. Furthermore, in response to the increased accessibility created by SFI's logging roads, external parties were manoeuvring to gain access to the timber on State Land forest. Most villagers were gripped by a sense of anxiety and apprehension.

Customary Claims and State Lands

The unusual shape of the Ulu Padas State Land/Forest Reserve boundary derives from the resource mapping process to define the Permanent Forest Estate in the then newly independent state of Sabah.[5] In this remote area, the Forestry Department relied heavily on aerial photographs to demarcate the boundary along signs of previous land clearing. The State Land boundaries delimit the area in which natives can apply for title under the *Sabah Land Ordinance* 1930 (little changed from the original legislation drafted in the days of the British North Borneo Company). Today, partly owing to the low-impact nature of traditional swidden cultivation, a substantial portion of the State Land still retains excellent forest cover, particularly on hill slopes and along the rivers. With the settlements and farms now located predominantly in the northern half of the 12 300 hectares, the southern section (about 60 per cent of the total area) has reverted to mature secondary forest. Although this area contains evidence of previous longhouse settlements, it now seems to play a more general function as a forest preserve for Long Pasia. Locals use longboats to access this area for fishing, hunting and resource gathering, and its importance has increased in light of the anticipated exploitation of the Forest Reserves. Because of its

---

[5] Sabah ceased to be a territory of the British North Borneo Company when it became part of Malaysia in 1963. The subsequent forest inventories conducted in the early 1970s and again in the late 1980s have been said to have accelerated the depletion of forest by providing a veritable 'treasure map' of the timber resource (Mannan and Awang 1997: 7).

impressive scenic and historic assets, this area is also the focus of tourism activities initiated with the assistance of WWF-Malaysia.

Insecurity of Tenure and External Threats

Although local people generally perceive that farms, fallows, homesteads and what has been a traditional forest resource (often loosely referred to as 'kampung land') is 'theirs', the Land Ordinance 1930 states that the land and forest of this area continues to be vested in the government until such time as it is administratively classified as Native Title or some other provision under the ordinance. The British North Borneo Company, which was the architect of this land legislation, clearly intended for local people to have secure tenure over their lands to facilitate its productive and commercial use (Singh 2000: 241). It enshrined the right for any individual who is a native of Sabah to apply for Native Title over a maximum area of 20 acres. Today, most people hold official Land Application receipts for the claims that they have filed with the Department of Lands and Surveys, but nothing is truly secure until the title is in their hands. The department has the formidable task of deciding on the legitimacy of claims throughout Sabah, resolving conflicts, and surveying the land. Not surprisingly, applications typically take decades to process and approve, especially in the more remote areas. This contributes to a high level of impatience and frustration at the local level.

> Whenever we visit the government offices, they tell us that the forest belongs to the government, and that we have no rights to the land of our ancestors. They say that if we want land we just have to apply for titles, but we've already done that years ago. Yet we are still waiting! (Long Pasia man at community meeting, 20 April 1999).

An area the size of the Ulu Padas State Land, with substantial forest cover, inevitably becomes the target of keen interest by external parties desiring to acquire forested land. In addition, without formal recognition of Lundayeh ownership of hinterland resources, there is no mandate for local people to exercise stewardship of these resources. From the mid-1990s, logging roads had already made the area accessible to recreational hunters and logging camps, and rivers were being fished by unsustainable means such as electricity and poison. At face value and from a conservation perspective, assisting the local community in securing ownership of this area was considered one way to exclude external logging interests, place the area under some form of communal management and include some provision for conservation. In addition to containing good samples of contiguous riverine oak chestnut and Agathis forest, the Ulu Padas State Land also contains several patches of rare kerangas forest. The longer the process of securing tenure was delayed, the greater the likelihood that logging contractors would obtain Temporary Occupation Licences to log the State Land through unscrupulous manoeuvrings of their own.

Seeking Conservation Through Land Tenure Security

Upon cursory examination, the bureaucratic processes and rigid criteria related to land application appeared to be a significant obstacle. Land legislation had a tendency to mystify local people who have only partial understanding of the options available to them to secure both ancestral farmland and forest. In Ulu Padas, these difficulties were accentuated by the fact that the amount of land available to local people has been reduced, with sizeable areas now classified as government-owned and designated for commercial purposes. There were several other problematic aspects that required solutions, namely the strong bias towards the conversion of forest to agricultural use and the scant provision made for landholding institutions that would support traditional agriculture[6] and maintain communally-owned forest reserves.

WWF-Malaysia's work with the community was founded on the belief that providing advocacy and mediation between the local community and the state could bring improved security of tenure and an opportunity to defend the forest. With this objective in mind, time was devoted to obtaining a clearer understanding of customary claims to land and investigating ways to translate these collective claims into a format supported by the *Land Ordinance*. The underlying intention was to mediate the process of communicating tenure claims in ways that would be accepted by the government system.

In principle, while some would argue that the conversion of traditional rights into colonial terms oversimplifies the original fluid nature of traditional land use, in the interest of expediency it was clear that government officers could more readily work with proposals that were supported by existing legislation. With imminent threats facing this area, expediency was preferred to the pursuit of an ideological crusade for indigenous rights. In many ways this action seemed to be supported by the fact that local people themselves deferred to the authority of government, and used their understanding of the *Land Ordinance* (however rudimentary) in their interactions with government agencies. Therefore the approach was not altogether inconsistent with local people's own acceptance of the legal framework of government.

Unfortunately, the initial assumption that assisting local people in securing tenure would be a straightforward matter of compiling a clear representation of customary tenure with which to seek the indulgence of government, proved to be naïve. We were soon to learn that customary claims were in fact a hotbed of

---

[6] The *Land Ordinance* emphasises that all titled land be put immediately to productive use, leaving little provision for recognising the need for farmers to have sufficient land to accommodate swidden rotations and fallow land to ensure long-term productivity.

contention. It was actually at village level that the full spectrum of conflict and irresolution emerged. Local people's claims were notoriously contradictory, with various factions competing for land within the 'community' itself. The multiple and divergent claims to land proffered by the local population seemed to be motivated by individual advantage rather than an adherence to time-honoured land tenure orthodoxies. Defining customary claims in ways that would satisfy all members of the local community presented many problems. This quarrelsome scenario is a consequence of the historic origins of the community, now more complicated by the influences of modernity and nascent self-interest.

The Difficulties of Defining Traditional Tenure

Traditionally, for the Lundayeh, in common with many other Borneo peoples (Appell 1986; Rousseau 1990), rights to a territory were held by a longhouse (Elmquist and Deegan 1974). Within this territory, any longhouse member could clear the forest to make a swidden. If an individual cleared a patch of forest with no known history of clearance, he and his descendants could lay claim to this land (Appell 1995). A hundred years ago, the Lundayeh settlement pattern in Ulu Padas was unlike that of the present day. The population was much larger and more widely dispersed in as many as nine longhouse hamlets (Hoare 2002: 31). Different longhouse groups had minimal interaction with each other, and clashes between them could be violent.

Prior to the arrival of Christian missionaries in the 1930s, the Lundayeh were one of Sabah's most feared headhunting societies. The advent of Christianity in the Ulu Padas uplands gradually eroded customs, traditions and beliefs, and tended to have a unifying influence. Over time, the different longhouse groups (now pacified) became more centralised, yet the apparent unity of the community was still undeniably undermined by age-old divisions carried over from the past.

In the period since then, there has also been substantial population movement to and from Ulu Padas. In the 1950s, government relocation programs encouraged the isolated population to settle in the lowland for greater access to amenities and services. In those days, the Ulu Padas villages were several days' walk from the end of the furthest dirt road. This meant great difficulty in accessing modern needs and markets for forest products. Children had to walk to their boarding schools in Sipitang. Many families opted to resettle in the new lowland villages. Nevertheless, others found it difficult to adjust and returned to re-establish the present-day villages (Hoare 2002: 35). Since then, there has been the usual population movement according to family circumstances. The relative porosity of the border has also enabled relatives or brides to come from other Lundayeh groups in Kalimantan. Many of these 'newcomers' have lived here for decades, becoming an integral part of the community.

Today's idyllic village of Long Pasia, with its central village leadership, church and school, is a relatively recent entity. Government centralisation policies to ensure more efficient administration (links to district government, agricultural schemes and subsidies), border security (army border scouts, immigration post) and provision of infrastructure and services (rural airport, clinic, primary school) have formalised the existence of the village as we know it. However, it is arguable that, despite appearances, and based on its disparate origins, the village might in spirit be more meaningfully viewed as several families cast together by history and circumstances. A superficial unity obscures the existence of enduring inter-family discord,[7] in addition to the usual feuds, disagreements and personality clashes that tend to colour village affairs. This has undermined the ability of the Ulu Padas 'community' to initiate collective action for common objectives and hindered the smooth resolution of tenure claims.

Advancing Claims: Exploiting Ambiguities in Interpretation

In this transitional period opportunities arise to exploit ambiguities and confusion in translating customary claims into legal title. As Peluso observes, the 'superimposition of statutory legal systems on customary systems creates new windows of opportunity for people to take advantage of multiple systems of claiming resources' (1995: 401). In the specific case of Ulu Padas, this has been a divisive process in which some groups have sought to boost individual advantage at the expense of others. While the State Land area of 12 300 hectares might be considered sufficiently large for a population of just over 500, local people had yet to come to a consensus on how customary claims might be realigned to fit the land now allocated to them. The somewhat arbitrary boundaries drawn to differentiate Forest Reserve from State Land excluded large areas (more than 3000 hectares) encompassing the customary land of some of the family groups of the Ulu Padas community. Most of the land in the vicinity of the present village centre is claimed by a handful of families under the traditional system of ancestral land clearance. A narrow interpretation of customary rights would advantage those with claims within the State Land while disadvantaging those without.

The Department of Lands and Surveys' Ulu Padas files comprise a tangle of separate land application approaches spanning many years. Multiple individual and joint Land Applications of various sizes have been filed, many overlapping with each other several times over. In order to plump up the size of the claim, a common strategy has been to produce a long list of joint claimants. In addition, a great many claims are being made by urban Lundayeh who may have had an ancestor from this area but have no real connection with the area at the present time. Other claims are being made for land by non-Lundayeh, utilising the

---

[7] Local people seldom discuss previous warfare as it is considered part of a 'shameful' past.

provision that allows Sabah citizens to apply for Country Leases for land development (or speculative) purposes.

There are contrasting points of view amongst the Lundayeh themselves over who has the right to claim land. For example, descendants of Lundayeh who resettled in other villages and towns maintain that they have legitimate ancestral claims to land. The 'founding families' that form the nucleus of the revived Ulu Padas villages believe that they have a greater claim as they returned and rebuilt the settlements through considerable hardship. They perceive the previous groups that left the village as having no claim under native customary law. People who leave their lands technically relinquish ancestral claims following a usufructuary principle in which land reverts to the community if its owners abandon the area. This is meant to optimise the allocation of land and resources to contemporary needs and current residents. Some dominant families are not prepared to reduce their claims to State Land to accommodate other families whose customary land does not fall within the designated State Land area. Finally, 'newcomers' (those that have either come from Indonesia or have returned to Ulu Padas from elsewhere over the past 25 years) are viewed by some as having no valid claim to land at all. For these families, the only option is to rent, borrow or buy land.

Although some applicants are clearly making excessive claims, without a clear and widely accepted understanding from the community of what a legitimate claim is, who legitimate claimants might be, or at least criteria to prioritise claims, it is not clear how a government land officer should begin. Instead of undertaking a joint initiative, different households or family groups were each pursuing applications separately. For some, this was a deliberate manoeuvre to exploit the lack of clarity regarding native customary rights to advantage their claims.

The jostling for advantage in the race to secure land tenure in this case calls instead for a combination of wealth, stature, and useful connections. Finding ways to lubricate the process and establish links with people in positions of influence has become a particular focus of people's efforts. Regrettably, the bureaucratic government system in Sabah can be, and has been, subverted on occasion. Lacking confidence in the fairness of the system, local people have become convinced of the need to assure outcomes through more deliberate means.

Divided They Fall

External parties interested in logging forest on State land are only too willing to offer their assistance. A common strategy has involved 'outside investors' using their connections to speed up processing of their local partner's Land Application in exchange for permission to apply for a Temporary Occupation Licence needed to conduct logging operations on State Land. Once the expensive surveys are carried out and the timber is removed, the land reverts to the Native Title holder. This procedure is potentially attractive to someone frustrated by the slow legal application process and eager to obtain a personal share of the proceeds from the sale of timber.

Participatory problem-analysis sessions conducted during the project inception phase in 1999 found the community to be chronically divided on tenure issues. Although there was genuine support for conserving forest resources and preserving the Lundayeh lifestyle, it was clear that certain parties were impatient to profit more directly from timber. It was common knowledge who the 'dealmakers' were and yet, to maintain appearances, the same individuals often railed openly against the evils of logging at public discussions. Adopting the emotive rhetoric of ancestral rights and dependence on the forest, they were concurrently pursuing their own projects such as securing road extensions to their farms, expanding cash-cropping orchards, and arranging for logging companies to operate on their land. To counter this, others claimed to be trying to secure large land areas through similar means, mainly to defend the communal forests from the destructive agenda of their neighbours.

Without a strong central leadership, there seemed to be an inability to mobilise a progressive course of collective action. It was becoming increasingly apparent that not everyone was being upfront about their plans and motivations. As 'deal-making' was perceived as being widespread and uncontrolled, more local people became convinced that they too needed to strive to get what they could while they could. Faced with this troubling scenario, many villagers conveyed their hopes that WWF-Malaysia, as an external entity, would take on the complex and uncomfortable task of ensuring the equitable distribution of land and conservation of communal forest. It was clear that a far more elaborate tenure solution was called for. Simply advocating the wholesale adoption of 'traditional' claims, even if such a thing could be defined, would likely lead to outcomes that neither supported wise resource management nor assured long-term community welfare.

The Community–Conservation Link

Misplaced Confidence

In the course of working with the villages, several unfortunate events illustrated the problem of too easily drawing a causal link between strengthening local claims and safeguarding natural resources. Initially, the advocacy strategy of defending local welfare and rights to resources proved surprisingly successful, albeit on a small scale. In mid-1999, community appeals to stop a logging contractor from logging an area of communal forest near the village of Long Mio garnered unprecedented media attention. The multi-agency taskforce appointed by the Forestry Department to seek a solution to this conflict (*The Star*, 9 May

1999; *Daily Express*, 10 May 1999) deemed that since the community objected to logging in this area, it would be left to them to negotiate terms with the contractor.

This was a tentative victory for the village: tentative because less than a year after the community had historically turned the contractor away, logging in this area resumed. Evidently, suitable terms for logging to resume *had* been negotiated by the headman without consulting with other community members. He argued that it was within his power to make the deal since the land involved was under his customary claim. Ironically, in 1997 the same individual had implored WWF-Malaysia to help prevent logging in this area. Now he was challenging us to make an attractive offer to conserve this forest 'since we were so keen on it'.[8] We had nothing to offer, except perhaps the wry reiteration that we had been assisting under the impression that it was the desire of the community to conserve this forest because it was of value to them, and not because of the prospect of inducements from us.

A similar event happened in Long Pasia shortly afterwards. We were told that the same logging contractor had *mistakenly* crossed over an area of privately owned land and logged part of a forested hill inside the catchment area of the village's gravity-feed system.[9] The logging company paid some compensation to the landowner and the village and was given permission to remove the felled logs. What made this incident suspicious was the swiftness with which compensation for this incursion was organised. It seemed as though this scenario had been devised to shield local counterparts from appearing complicit in an *arrangement*. The incident was not reported to the Forestry Department. It was qualified that: 'If the Forestry Department comes, they only fine the contractor or the logs are confiscated. This way at least we get something.'

These two events suggest that the effectiveness of the community–conservation NGO partnership at raising awareness and sympathy for biodiversity conservation can be highly effective, but it can backfire quite easily. While it is possible for advocacy strategies to 'protect' local people's interests from outside threats, it cannot easily protect local people from themselves. Indeed, such strategies may quite inadvertently raise the rates of compensation and enhance the temptation to cash in for short-term gains.

Communal Resource Management: Ideals Versus Reality

It has often been argued that communal management of natural resources engenders greater social justice and preservation of the environment. Communal

[8] Other Long Mio residents attribute the headman's change of heart to his advanced age and inability to understand the long-term purpose of conserving these resources, as well as the attractive inducements from the logging contractor which provided more tangible benefits in the short term.
[9] For most rural villages in Sabah, such gravity-feed systems provide piped water from dammed streams to village households and farms. It is the only source of water other than the river and rainwater.

management has been portrayed as contributing to the sustainable use of natural resources and providing for local needs by ensuring the equitable distribution of land and resources. Further to these requirements, a functioning communal management system should be supported by a strong community organisation to arbitrate norms and regulations involved in managing resources held in common. On close examination, I have found that none of these three elements can be said to be truly functioning in the Ulu Padas villages at this time. Some may argue that this situation has arisen because local authority over customary lands and resources has been undermined in recent decades. It is also possible that in the right policy environment, all three of these prerequisites for strong communal management could be revived. However, the present situation does not engender confidence in the capacity of local communities to assume ultimate management of these resources. A very significant factor in this observation is the degree to which village life has been impacted by the pervasive influence of modernity. This is most apparent from some of the specific changes affecting common resources shared by the village community.

In Ulu Padas, traditional guidelines exist to govern access to resources that are held in common. A civil contract allows community members to access resources for domestic use both from each other's fallow fields and in the village's wider 'territory' according to stipulated regulations. Today, many common property regulations are not being effectively enforced and are openly flouted by some. When outsiders come into an area, they are customarily expected to ask the village headman for permission to enter the forest to harvest plant resources, go hunting or fishing. However, today this is often ignored. Consequently, it has become increasingly difficult to control the unsustainable exploitation of resources. In the rivers and streams, forbidden poisons and electric current have been used. Recreational hunters from urban areas are now using logging roads to access hunting areas (reports of six or seven deer and wild boar taken in a night are common).

Although the 'enemy' is frequently characterised as the evil outsider, often entry is facilitated from within. It is common for local people to serve as paid guides on these fishing and hunting excursions, and some even use unsustainable fishing practices themselves. Logging camps in the uplands create a steady demand for wild meat and this is a prime source of income for village hunters. This commercialisation of wild game already represents a form of open access use as it is contrary to conventions that restrict use of the resource to domestic needs (Berkes et al. 1989). With money now an important motivation, detractors who have psychologically crossed out of the traditional paradigm are unconcerned by social sanctions against such practices. While the removal of local people's authority to exclude outsiders is a consequence of state laws, it is inconclusive whether this is the sole cause of the erosion of local management systems.

Local response systems for ensuring the smooth working of commons management were also not actively functioning. In community consultations, the women's and young people's discussion groups complained that irresponsible cutting of timber in nearby amenity forests was reducing the supply of accessible firewood, thereby burdening them with the need to travel further to replenish the hearth. Indiscriminate clearing of land along upstream riverbanks was also silting up patches of wild vegetables that are collected for daily meals. Traditional systems were not actively addressing resource use conflicts or regulating the activities of fellow community members. In addition, there seemed to be no framework for women to raise their specific concerns (Vaz 1999: 5).

A further development has been the strong trend towards privatisation of all resources, despite there being a long tradition of community access to certain resources such as bamboo shoots, fruits and others. Although this is the cause of considerable ill feeling, such behaviour being seen as mean and not customary for the Lundayeh, it has not been openly objected to. Rather, it has led to other people following suit in cordoning off other resource areas (Hoare 2002: 35). Increasingly, there is also a trend towards asserting exclusive use of all land. In the past, fallowed swidden land would traditionally be loaned to kin or neighbours for farming if needed. There is a new emphasis on the need to use land commercially for permanent crops and to secure this land with heavy emphasis on the principle of inheritance based on descent. In this context social obligations are being downplayed. Commercial crops are being emphasised in order to generate cash incomes (ibid.: 172). This can also be said to reflect a strategy to to strengthen the perceived legitimacy of land claims with the investment of labour on developing permanent crops, which would be viewed as being 'more progressive' by government authorities.

Moving Beyond the Impasse: Teasing Out a Tenure Solution

In Ulu Padas, it was clear that in a leave-alone scenario, the villagers would be unlikely to *automatically* assert forms of management that would necessarily uphold environmental conservation and equitable access to land and communal resources. If land in Ulu Padas were to be awarded on the basis of ancestral claims alone, certain individuals or families would lay claim to vast tracts of land, more than they could feasibly use for agriculture, while others would have no such claims despite their having lived in this area for 20 years or more. With the prevailing trend towards privatisation, there was no guarantee that the former customary system of loaning land to fellow community members would be honoured. And with so many large claims focused on the forested area in the southern half of the State Land, applicants hoping to make their fortune through timber deals could deprive the larger community of vital shared resources while precipitating serious environmental degradation.

From a conservation perspective, any large claim to land awarded to individuals or select groups without provisos for accountability to the total community would expose it to unsustainable exploitation and negative environmental impact. This is true regardless of whether the applicants are *bona fide* community members residing in the village, Lundayeh people who have moved away from the area, or well-connected Sabahans seeking to obtain land for development. If anyone were to be given ownership of a large forested area, the individual would be able act independently of community interests. The lure of selling rights of access to timber *taukeh* (tycoons) would be too difficult to resist.

As far as WWF-Malaysia's objectives were concerned, in identifying a common standpoint from which to evaluate resource tenure solutions, it was necessary to outline a clear set of principles with which to uphold requests to protect local livelihoods, cultural heritage and the living environment that had been voiced by the community in earlier discussions. Bearing in mind the organisation's core business, the tenure solution would also have to support biodiversity and environmental conservation. In addition, ways would have to be found to support local access to communal resources and to restrict external interests. Ultimately, any intervention would have to promote the equitable division of land to all Ulu Padas residents and ensure that the activities of a few do not have the propensity to disadvantage the larger community.

Playing the initial role of a go-between, the WWF-Malaysia project officers consulted with the district officers of the Lands and Surveys Department and other government agencies to better understand the official process by which native land claims could be resolved, the specific provisions for native tenure (both individual and collective) within the *Land Ordinance*, and the legislative procedures by which local communities might formalise claims for land. The community's confusion over the complicated and confusing process was communicated, while the government officer clarified some of the obstacles and limitations hindering the smooth and speedy resolution of tenure from the Department's perspective. These difficulties were a common concern of both parties; after all, it is technically in the Department's interest to find expedient means to complete the statewide land-registration process.

Relevant legislation was translated or explained in Bahasa Malaysia in order to familiarise community leaders and organisers with the land application process and supporting legislation. Local people became quite proficient at interpreting laws and policies governing environmental protection, sustainable forest management and native land tenure. With an ongoing dialogue established with the District Surveyor, in a matter of weeks what first appeared to be an intimidating and impenetrable bureaucracy evolved into a joint strategy. The District Surveyor was exemplary in upholding the spirit of the *Land Ordinance* and the government's original intention that the Ulu Padas State Land should be entrusted to the Lundayeh people, suggesting the most effective routes towards this objective.

Many members of the community still believed that it was possible to lay claim to the entire Ulu Padas region, including the Forest Reserves. Hearing directly from a government officer that this was highly improbable helped local people to abandon unrealistic expectations and reorient them towards more achievable aims sufficient for their needs that could be endorsed by government agencies in accordance with current laws.

Individual Titles

In order to assure that local residents were given top priority in receiving individual land titles, the Department of Lands and Surveys first began a process of filtering the volumes of applications on file to prioritise families with a recognised claim and need. Village leaders and committees provided a vetted list of names to facilitate this. Second, in order to avoid the obstacles of overlapping claims, plans were made for all Native Titles in the Ulu Padas State Land to be processed in one block. At a future date, Lands and Surveys officers would base themselves in the village for a time to consult with the community to demarcate the location of household plots to be awarded Native Titles close to the main village centre and most active agricultural areas. Similar approaches have already been used with considerable success in the adjacent Beaufort district. In this way, the processing time and survey costs would be greatly reduced.

The Sabah Biodiversity Conservation Project soil and slope studies were consulted in determining the distribution of fertile land suitable for permanent crops and less fertile land for mixed cultivation. The Lands and Survey Department had already earmarked slopes and catchments that would automatically be reserved as amenity forests for domestic use. With invalid or less valid claims removed or reduced in size, the potential threat of alienating large areas of forest to external parties was mitigated.

Safeguarding Communal Forest

Care was taken to ensure that the move to proprietary rights would not undermine the importance of shared forest resources (Li 1996; Stevens 1997). Without access to the Forest Reserves that had served as a wider resource hinterland, it was imperative that a reserve be established within the State Land to safeguard resources for domestic access. Since the Native Title provision applies only to smaller parcels of land intended for productive use, a Native Reserve was the best means by which a large contiguous area of forest could be protected while still enabling local use by the village as a whole.[10] Use rights would only be extended to villagers and guidelines for harvesting resources would be determined so that each community member would have an equal role in ensuring the appropriate use and management of the area. Here then was the possibility to re-establish a secure resource from which common property regulations could be negotiated anew.

An application for a Native Reserve of 4500 hectares in the forested southern section of the State Land was submitted by Long Pasia in October 1999. This area incorporates the former settlements, burial sites, rock and earth monuments and historical routes, including numerous sites of value for biodiversity and tourism development. As the face of Ulu Padas begins to change, the proposed reserve is intended to protect at least some of the most cherished elements of the Lundayeh lifestyle and identity before they are lost.

However, the process of obtaining the endorsement of all members of the community for the Native Reserve was often frustrating for the community members striving to put the proposal together. Certain segments of the community were antagonistic as this form of shared tenure would upset private timber deals. At times it seemed that the signed endorsement required for the Native Reserve proposal would never be secured. The application languished uncompleted for two months until, in September 1999, information had filtered in from several sources that a logging company with an influential former politician as its director was close to being awarded this area for logging. This confirmed earlier warnings of the imminent threats from external interests. A collective application for a Native Reserve was swiftly formalised and submitted to the government by villagers of Long Pasia. Local people were certainly not going to let an outsider's claim usurp theirs. Long Mio followed suit, proposing another area of several hundred hectares as Native Reserve.

The Native Reserve applications have now been prioritised on the merits of the communal claim, causing other land applications for the same area to be rejected. The Native Reserve application has already been approved at several levels and is now in the final stages of processing. If it makes it through the final stages, the Ulu Padas Native Reserve could be one of the largest areas of communal forest to be established in Sabah in recent history.[11] However, the hesitant steps taken towards its establishment suggest that aside from its ability to neutralise outside threats, gazetting a Native Reserve will not in itself guarantee conservation outcomes.

---

[10] All other 'protected area' legislation such as those used for the establishment of Sabah Parks or Wildlife Sanctuaries explicitly forbids access and use of the area concerned by local people. This has understandably nurtured a natural opposition to protected area proposals.
[11] An 'if' still applies as policies and leadership tend to change frequently in Sabah. The conservation quest is littered with premature stories of victory, followed by bitter disappointment.

Strengthening Communal Resource Management

With the initial obstacle of tenure insecurity overcome, the Ulu Padas community still faces the important challenge of re-establishing its communal resource management systems and institutions. The task has barely begun and will need a commitment of resources and external support if it is to be successful in the long term. Government agencies and NGOs have a role to play in guiding catchment protection, biodiversity conservation, and the management of tourism and recreational areas. Discussions need to be held to elaborate management and use guidelines for the Native Reserves as well as provisions to ensure the necessary levels of accountability and transparency in the management of this important area. Bearing in mind the stratified nature of most communities, care must be taken to ensure that decision making in the name of the community is not usurped by more powerful elements within it. If such intra-community equity in decision making is not assured, even participatory modes of resource management would fail to deliver equitability (Singh et al. 2000).

The Importance of Collaboration

The Role of Policy Reform in Reversing Destructive Trends

In Ulu Padas we observed how community institutions have become weakened by the absence of tenure security and the impacts of monetisation and opportunism. Anecdotal evidence suggests that similar scenarios are being replayed in rural communities throughout Sabah. The uncertainty surrounding land alienation, and the potential profitability of making claim to and selling timber rights, manifests in actions that are deleterious to the welfare of local communities, to the environment and to long-term development. As communal forest areas continue to come under threat, divisions within communities are precipitated by outside interests to undermine their defence of shared resources. At this crucial juncture, the state government has the potential to intervene to reverse these trends.

There are some immediate steps that can be taken to improve current policies and practices governing the management of forests and land use change in Sabah, in particular those that are inadvertently encouraging resource degradation such as the policy of handling land applications on a piecemeal basis. To support the social integrity of these communities, the land registration process should be conducted by engaging village communities as a whole. Village land use and future development plans should be mapped out, and designated sensitive areas and common property resource areas identified with the mediation and supervision of officers of the Department of Lands and Surveys, the Native Court and other observers. In addition, the integrity of these plans should be upheld by all government agencies that have the authority to issue logging or occupation licences so as to restrict activities that threaten resource management in the community area.

Working with the State

The experience in Ulu Padas also demonstrated the value of collaboration in conserving communal forests. Government officers, conservation practitioners, researchers and scientists, and, of course, local people have the potential to complement and reinforce each other's contributions. Accounts that portray the contest over resources as lopsided battles between state élites and marginalised communities have an obvious emotional appeal, yet they can dangerously polarise issues. I have found that the state government includes people who are receptive and committed to conservation and community interests, and who try, within their limited mandates, to seek favourable outcomes.

Working with local people has given me an appreciation for their resourcefulness and eagerness to be engaged more actively in developing their economic potential and building stronger futures for their families, while retaining links with their land, identity and heritage. Most perceive these aspirations as being achieved through opportunities arising from inclusion in state development programs, such as support for agriculture and, more recently, nature tourism. These impressions resonate strongly with Li's observation that, 'supporters of peasant struggles who assume that "traditional" communities are inclined to oppose "the state" in order to preserve "their own" institutions and practices may overlook the extent to which uplanders seek the benefits of a fuller citizenship' (Li 1999: 21). There is no question that the residents of Long Pasia and Long Mio see development in terms of fuller integration in the state system and through government-funded infrastructure and services.

Using a 'practical political economy' mindset (Chambers 1983), working more closely within and through the system in Sabah has enabled conservation NGOs to have continued access to relevant spaces, be they actual physical locations or the opportunity to provide input on important issues. Non-governmental organisations that 'act responsibly' are in a better position to increase the credibility of local-level conservation initiatives and maintain an opportunity for continued advocacy. Accordingly, the bid to secure communal ownership of forest resources and Native Title received a favourable response as the strategy adopted was consistent with the land legislation and was pursued through the official channels.

Discussion: Conservation on Community Lands

Communal Lands as Spaces for Biodiversity

Across the globe, the spaces reserved for biodiversity conservation are decreasing dramatically (Cox and Elmquist 1991: 317), and as Stevens (1997) points out, very few wilderness areas can be considered uninhabited. Thus it is inevitable that there is a convergence of interest in community-claimed lands for conservation. In Sabah, local communities that occupy the last spaces where Borneo's biologically rich rainforests persist can be important agents in the quest to ensure its continued existence. Native communities have customary claim to some of the state's fragmented natural areas, and many are deeply concerned about the environmental impact of logging and land clearing. Politically, those whose next meal or glass of water will come directly from threatened environments are naturally perceived as having a greater moral right to defend their livelihoods and living environments. Indirectly they stand to be a voice for forest conservation.

In the case of Ulu Padas, the attention garnered by the community's campaign to defend communal forest areas also cast a spotlight on logging operations in the surrounding Forest Reserves. With mounting criticism of logging in highland areas from the public, there was increased pressure on government agencies to control environmental damage and conserve important areas. The Forestry Department was able to extract a greater commitment to Sustainable Forest Management principles from the concessionaire — including setting aside areas with steep slopes, important biodiversity areas, wildlife corridors and areas important to the community, including identified tourism development sites. In addition, a large area designated for Industrial Tree Plantations has now been reassigned to Natural Forest Management. The Department of Environmental Conservation was empowered to play a stronger role in enforcing environmental regulations. The interest in the area for tourism also generated discussions of collaboration to develop alternative economic activities in the Upper Padas. In addition, there has also been gradual progress in discussions to establish a transboundary conservation area with Indonesia and Sarawak. Arguably, many of these developments would not have taken place had the community not played a role in calling attention to the threats to the environment in this remote corner of the state.

Although resolving tenure in community-claimed lands can be an exceedingly complicated undertaking, there is immediate value in arresting the divisive competition for land and forests in which conservation, communities and resource management are all losers. Concluding the period of ambiguous transitional tenure has the potential to provide an improved foundation for the future, and an impetus for the community to heal and come to terms with a new set of circumstances. Restoring stewardship of forested land to local communities may yet be a promising means of achieving conservation goals. As Sabah's forest heritage continues to be whittled away by a combination of both human and natural agents, in years to come such preserves could become exceedingly important as refuges for what remains of wild Borneo.

Making Community–Conservation Partnerships Work

The Ulu Padas case study was an experimental partnership between a conservation NGO and a rural indigenous community with a focus on strengthening local tenure arrangements. A generally symbiotic relationship was struck between the two parties — each motivated by a specific payoff. Local communities marginalised by the complicated legal procedures for land ownership leaned on the influence and expertise of an established organisation to assist in securing property rights and stimulating tourism initiatives. By providing the Ulu Padas community with information, legal advice and access to government channels, strategies for obtaining tenure security were expedited, the alienation of land and resources from local people was challenged, and the loss of biodiversity was mitigated.

However, in such partnerships, conservation NGOs are potentially at the mercy of communities. Non-government organisations do not have the authority to *impose* their will since it is local people that have claim to the land. The conventional methods are to inform, persuade, and sometimes develop livelihood alternatives or provide monetary incentives. Inevitably, it is the community that has to make the final decision and this implies a fairly high risk of failure. The experience in Ulu Padas has illustrated that, the close similarity of goals notwithstanding, collaborations between communities and conservation NGOs, however cordial, would be better treated as business partnerships built on compromise, not assumptions of mutual goodwill and shared objectives. In reality, each party gives up an ideal in order to achieve a reduced benefit. Not quite a win–win scenario, but perhaps the next best option.

In the case of the community, people must be reconciled with the sacrifice of short-term gains in order to achieve long-term resource security and some development assistance. In the case of the conservation organisation, domestic use of forest is supported (or tolerated) in order to achieve specific biodiversity-conservation objectives. The two parties are thrown together by mutual need because external threats would be impossible to repel independently. However, should the terms of this agreement be contravened in any way, the partnership becomes meaningless.

Although I believe that it is still important to be open to the possible contribution of local communities and communal areas to biodiversity conservation, it is necessary to concede that this should not be equated with 'absolute, unmediated and entirely unregulated control over biodiversity resources' (Singh et al. 2000: 72). The strategy of safeguarding customary tenure does not automatically beget conservation outcomes. For this reason, conservation practitioners have advocated that conservation objectives be explicitly spelled out through the use of Negotiated Contractual Agreements. 'Essentially this involves the state or the official conservation agency negotiating with the local communities and coming to an agreement on their rights and obligations regarding the conservation of bio-diversity or natural resources in their area' (ibid.: 74). Such a process requires the clear articulation of each other's commitments and responsibilities.

Conservation NGOs need to be upfront about their own agenda, and recognise the fact that they cannot operate out of altruism alone. They have to answer to donors and justify how project activities will contribute to a specific 'global mission'. They also have limitations: most notably in terms of funds, staff, economic expertise, and of course decision-making power. These aspects should be made clear to local people at the earliest possible stage, lest incorrect assumptions lead to disillusionment and misunderstanding.

In seeking to conserve biodiversity on communal lands, it is important to acknowledge that local people are being asked to bear the bulk of the burden of conservation in terms of social and economic impact (Wilshusen et al. 2002; Wells 1995). This may be understood in terms of restricted access to land and resources, or the opportunity costs of forgoing their exploitation. Local people are compelled to conserve and manage resources by an obvious hierarchy of motivations. Although religious or ethical imperatives, the availability of natural resources, the provision of ecological services, and fulfilment of aesthetic and recreational needs are important factors, direct and immediate financial returns are the most prominent motivation for most people (Singh et al. 2000).

Although there is a clear relationship between sustaining communal forest and the quality of the living environment, local people's aspirations usually extend beyond mere settlement and subsistence. It is imperative that program developers and policy makers accept from an early stage that '[u]pland populations have different degrees of attachment to their current locales and different degrees of commitment to an agrarian future' (Li 1999: 34). It remains an important question to ask whether conservation is a choice local people can afford to make.

In Ulu Padas, without sufficient financial backing for development alternatives, most local people saw WWF-Malaysia's assistance as well-intentioned but not pragmatic enough to address immediate economic concerns. At this stage there is still an opportunity to create a framework by which community development and biodiversity-conservation efforts might be mutually supportive. Economic incentives and support for income diversification strategies will need to be considered as part of any effort to conserve biodiversity. Fundamentally, unless conserving forest resources is *immediately and directly relevant* to supporting the livelihoods of local people, and is included in plans for development, the impetus for biodiversity-conservation outcomes may not be sustained.

Conclusion

In examining the contest for the Upper Padas forests, a series of philosophical questions emerges. Should any one party automatically be privileged over the other in the claim to forest? Do local people's rights to exhaust their own resources, to achieve their own priorities and short-term goals, supersede the role of government in marshalling the use of state resources for development? Are conservationists justified in valuing threatened biodiversity over either of these aims? These are questions without any easy answers, but they will need to be addressed as different interests increasingly contend for the scattered forest refuges that now remain. The most agreeable solution is for each party to acknowledge the interests of the others in order to come to a synergistic solution. The danger for conservation organisations is that it is often too easy to over-extend assistance and to take on the concerns of communities without assuring that biodiversity conservation retains its primacy.

I believe that communities should assume some responsibility for biodiversity which is in their care. However, governments with whom nations vest this important duty have the most prominent role to play in protecting important areas and supporting compatible economic activities in such areas. The economic opportunity cost to communities will need to be considered. Using the concept of negotiated agreements, economic assistance could be developed and incorporated by governments and NGOs into agreement packages, with the understanding that the benefits will be withdrawn if the substance of the agreement is violated. Ultimately, as long as communal lands continue to act as *de facto* refuges for threatened biodiversity, there is a role for conservation NGOs to bring together the parties concerned to creatively secure the protection of these areas and improve development prospects for local communities. Policy reform, addressing tenure in collaboration with government partners, and tackling economic issues are some of the key elements to such a strategy.

References

Appell, G.N., 1986. 'Kayan Land Tenure.' *Borneo Research Bulletin* 18: 119–130.

————, 1995. 'Community Resources in Borneo: Failure of the Concept of Common Property and its Implications for the Conservation of Forest Resources and the Protection of Indigenous Land Rights.' In G. Dicum (ed.), *Local Heritage in the Changing Tropics: Innovative Strategies for Natural Resource Management and Control* . New Haven (CT): Yale University.

Baland, J. and J. Platteau, 1999. 'The Ambiguous Impact of Inequality on Local Resource Management.' *World Development* 27: 773–788.

Baviskar, A., 1995. *In the Belly of the River: Tribal Conflicts over Development in the Narmada Valley*. Delhi: Oxford University Press.

Berkes, F., D. Feeny, B.J McCay and C.M Acheson, 1989. 'The Benefits of the Commons.' *Nature* 340: 91–93.

Brookfield. H., L. Potter and Y. Byron, 1995. *In Place of the Forest: Environmental and Socio-economic Transformation in Borneo and the Eastern Malay Peninsula*. Tokyo: United Nations University Press.

Chambers, R., 1983. *Rural Development: Putting the Last First*. Harlow: Longman.

Colchester, M., 1997. 'Salvaging Nature: Indigenous Peoples and Protected Areas.' In K.B. Ghimire and M.P. Pimbert (eds), *Social Change and Conservation: Environmental Politics and Impacts on National Parks and Protected Areas*. London: Earthscan.

Cox, P.A. and T. Elmquist, 1991. 'Indigenous Control of Tropical Rainforest Reserves: An Alternative Strategy for Conservation.' *Ambio* 20: 317–321.

Elmquist, T. and J.L. Deegan, 1974. 'Community Fragmentation Among the Lun Bawang.' *Sarawak Museum Journal* 22(43): 229–247.

Hazebroek, H. and A. Kashim, 2000. *National Parks of Sarawak*. Kota Kinabalu: Natural History Publications.

Hoare, A.L., 2002. Cooking the Wild: The Role of the Lundayeh of the Ulu Padas (Sabah, Malaysia) in Managing Forest Foods and Shaping the Landscape. Canterbury: University of Kent (Ph.D. thesis).

Hong, E., 1987. *Natives of Sarawak: Survival in Borneo's Vanishing Forest*. Penang: Institut Masyarakat.

Kulip, J., G. Majawat and J. Kuluk, 2000. 'Medicinal and Other Useful Plants of the Lundayeh Community of Spipitang, Sabah, Malaysia.' *Journal of Tropical Forest Science* 12: 810–816.

Li, T.M., 1996. 'Images of Community: Discourse and Strategy in Property Relations.' *Development and Change* 27: 501–527.

———, 1999. 'Marginality, Power and Production: Analysing Upland Transformations.' In T.M. Li (ed.), *Transforming the Indonesian Uplands*. Amsterdam: Harwood Academic Publishers.

———, 2002. 'Engaging Simplifications: Community-Based Resource Management, Market Processes and State Agendas in Upland Southeast Asia.' *World Development* 30: 265–283.

Mannan, S. and Y. Awang, 1997. 'Sustainable Forest Management in Sabah.' Kota Kinabalu: Sabah Forestry Department (unpublished seminar paper).

Paramanathan, S., 1998. 'Assessment of Soils, Ulu Padas: SBCP-IPPA Soils Assessment.' Kota Kinabalu: Ministry of Tourism and Environmental Development.

Payne, J. and J. Vaz, 1998. 'Ulu Padas — Final Report and Recommendations: SBCP-IPPA Technical Report.' Kota Kinabalu: Ministry of Tourism and Environmental Development.

Peluso, N.L., 1995. 'Whose Woods Are These? Counter-Mapping Forest Territories in Kalimantan, Indonesia.' *Antipode* 27: 383–406.

Phillips, A. and A. Lamb, 1997. 'The Botanical Richness of Ulu Padas: SBCP-IPPA Botanical Assessment.' Kota Kinabalu: Ministry of Tourism and Environmental Development.

Poffenberger, M. and B. McGean, 1993. *Upland Philippine Communities: Guardians of the Final Forest Frontiers.* Berkeley: University of California, Center for Southeast Asia Studies.

————, 1996. *Village Voices, Forest Choices: Joint Forest Management in India.* New Delhi: Oxford University Press.

Rousseau, J., 1990. *Central Borneo: Ethnic Identity and Social Life in a Stratified Society* . Oxford: Clarendon Press.

Singh, D.S.R., 2000. *The Making of Sabah 1865–1941: The Dynamics of Indigenous Society.* Kuala Lumpur: University of Malaya Press.

Singh, S., V. Sankaran, H. Mander and S. Worah, 2000. *Strengthening Conservation Cultures: Local Communities and Biodiversity Conservation.* Paris: United Nations Educational, Scientific and Cultural Organisation.

Sinun, W. and J. Suhaimi, 1997. 'A Hydrological and Geomorphological Assessment of Ulu Padas: SBCP-IPPA Hydrology Assessment.' Kota Kinabalu: Ministry of Tourism and Environmental Development.

Stevens, S., 1997. 'New Alliances for Conservation.' In S. Stevens (ed.), *Conservation Through Cultural Survival: Indigenous Peoples and Protected Areas.* Washington (DC): Island Press.

Towell, P., 1997. 'Conservation and Development in the Ulu Padas Area — an Analysis of Local People's Principles of Involvement: SBCP-IPPA Socio-Economic Assessment.' Kota Kinabalu: Ministry of Tourism and Environmental Development.

Varughese, G. and E. Ostrom, 2001. 'The Contested Role of Heterogeneity in Collective Action: Some Evidence from Community Forestry in Nepal.' *World Development* 29: 747–765.

Vaz, J., 1999. 'Ulu Padas: Assessment of Tourism Potential. ' Kota Kinabalu: WWF Malaysia.

Wells, M.P., 1995. 'Biodiversity Conservation and Local Development Aspirations: New Priorities for the 1990s.' In C.A. Perrings, K-G. Maler, C. Folke, C.S. Holling and B-O. Jansson (eds), *Biodiversity Conservation: Problems and Policies*. Dordrecht: Kluwer Academic Publishers.

Wilshusen, P.R., S.R. Brechin, C. Fortwangler and P. West, 2002. 'Reinventing a Square Wheel: A Critique of a Resurgent "Protection Paradigm" in International Biodiversity Conservation.' *Society and Natural Resources* 15: 17–40.

Chapter Eight

Social, Environmental and Legal Dimensions of *Adat* as an Instrument of Conservation in East Kalimantan

Cristina Eghenter

Introduction

Since the 1980s, there has been a radical shift in thinking about environmental and natural resource management as being inseparable from issues of the welfare and human rights of minority or indigenous people (Chartier and Sellato 1998). This view was also shared in conservationist circles, where indigenous people acquired increasing visibility in the management of protected areas. Indigenous people and conservation organisations came to be perceived as natural allies based on the evidence that:

> ... most of the remaining significant areas of high natural value on earth are inhabited by indigenous people. This testifies to the efficacy of indigenous resource management systems (WWF 1996: 3).

The preservation of biological diversity and natural resources was not only regarded as compatible with the rights and traditions of indigenous people, but instrumental to the efforts of many forest communities to protect their forest and defend their land (WWF 1996, 1998).

In this context, the adoption and application of local management practices and customary law is viewed as the key to success. The devolution of management responsibilities to local institutions and local leaders is based on the belief that these people are endowed with a natural capacity to manage a protected area in the interest of sustainability and biodiversity conservation. At the same time, however, little attention is paid to the historical, social and legal context of local institutions and custom, with little understanding of the premises that would sustain their effective integration into different political and legal regimes.

By drawing on the experience of an experiment in community-based management in the National Park of Kayan Mentarang, East Kalimantan, Indonesia, this chapter examines the ways in which customary regulations can be integrated with national laws with regard to the management and protection of natural resources. I focus my attention on customary law, or *adat*, as an ideological and ethical statement by the community with regard to criteria for resource management. I analyse the kind of resources regulated, and how they are regulated, as inscribed in local regulations among Kenyah and other communities inhabiting the area of the National Park. This is done in order to uncover potential points of intersection between criteria for natural resource management as practised locally, and the principles of management of protected areas contained in the documents of the Indonesian Government, and implemented by conservation organisations like the Worldwide Fund for Nature (WWF). I argue that uncovering potential points of convergence and difference is crucial to a productive engagement between law and custom, and the creation of future alternatives for effective and equitable strategies of conservation area management.

Adat Communities in the Kayan Mentarang National Park

Stretched along the mountainous interior of East Kalimantan, Indonesian Borneo, the Kayan Mentarang National Park lies at the border with Sarawak to the west and Sabah to the north. With its gazetted 1.4 million hectares, it is the largest protected area of rainforest in Borneo and one of the largest in Southeast Asia. A strict nature reserve since 1980, the area was declared a National Park by the Minister of Forestry in October 1996.

The history of the natural landscape of the park is inexorably intertwined with the history of its people. Extensive archaeological remains in the form of stone burials are found in the reserve. They date from about 300 years ago and were used for secondary burial rites.

About 16 000 Dayak people live inside or in close proximity to the Kayan Mentarang National Park. Roughly half of these people — mostly Kenyah but with a small number of Kayan, Saben and Punan — are primarily shifting cultivators. The rest — mostly Lun Dayeh and Lengilu in the north — are mainly wet-rice farmers. The inhabitants of the park and surrounding areas depend on hunting, fishing, and collecting wild plants for their subsistence needs. Trade in forest products such as gallstones (from langurs and porcupines) and aloes wood or *gaharu* (*Aquilaria* spp.), as well as revenues from temporary employment in Malaysia, are the principal ways to earn cash to purchase commercial goods, pay for school fees, and cover travel expenses to the lowlands. These activities have allowed them to meet their basic needs and be self-sufficient under stable circumstances. Average income levels of the people in many areas of the National Park are above the average level for the province of East Kalimantan. However, transportation costs are very high. Only the existence of price subsidies has managed to keep prices of essential goods under control. Nevertheless, local prices are still on average three to six times higher than in the lowlands. People in parts of Krayan, Lumbis and Apo Kayan often travel across the border to Malaysia to get sugar, salt and gasoline at lower prices.

Individual claims to land are established by cutting trees or clearing forest. The right to use the land is then passed on to the succeeding generation. Useful trees like fruit trees, illipe nut trees, cinnamon and honey trees are owned by individuals or kin groups. When people decide to move to another community, they transfer rights over fallow land and trees to family members or, in more recent times, sell the rights to other villagers.

Traditional forest areas with protection status or strict management regimes exist and are known among the different ethnic groups as *tana ulen*, *tana ang* and *tana imud*. *Tana ulen*, for example, is *tana* (land) which is *m/ulen* (restricted/prohibited). It is an expanse of primary forest rich in natural resources such as rattan (*Calamus* spp.), *sang* leaves (*Licuala* spp.), hardwood for construction (for example, *Dipterocarpus* spp., *Shorea* spp., *Quercus* spp.), fish and game — all of which have high use value for the local community. In the past, *tana ulen* functioned as forest reserves managed by the aristocratic families of the community. Nowadays, responsibilities for the management of the forest reserves have been transferred to the customary councils that oversee *tana ulen* forests on behalf of the entire community and according to customary law (Eghenter 2000a).

All other forest in the village territory may be regarded as community land with open access regulated by social norms. Although the boundaries of the territory are known to the communities, these are not enforced in an exclusive way. The model is akin to what Casimir defines as a 'social defense boundary' strategy (1992: 11–13) where there is no strong sense of territoriality and perimeter defence, and neighbouring people can access the area provided that they ask for permission.

The communities living in and around the park are still *adat* communities, largely regulated by customary law or *adat* in the conduct of their daily affairs and the management of natural resources in the customary territory or *wilayah adat* (Eghenter and Sellato 1999, 2003). The customary chief (*kepala adat*) administers the customary law with the help of the customary council (*lembaga adat*). All elected officials at village level and prominent leaders of the community sit on a customary council.

In the past, the customary chief acted as coordinator and decision maker on civil and religious matters concerning the groups in the territory under the chief's jurisdiction (*wilayah pemerintahan adat*). During the colonial period, the authority and jurisdiction of the chief were recognised, although the Dutch administration occasionally intervened to force changes to the boundaries of the territory or lend legitimacy to a particular group in case of dispute. It was only after independence, and the introduction of a new administrative system based on sub-districts (*kecamatan*), that the status of *wilayah pemerintahan adat* changed to become simply *wilayah adat* (customary territory). The *kepala adat* became a sub-district headman, working with the sub-district officer and receiving an honorarium from the government.

Notwithstanding the assimilation into government bureaucracy, the role of traditional institutions like *lembaga adat* is still key to understanding the communities' views and the way they deliberate on issues of forest and natural resource management. The results of a recent participatory inventory of local institutions, conducted by the Community Empowerment Team in the National Park area of the Kayan Mentarang Project (Lawai 2001), show that the *lembaga adat* was used more for dealing with local affairs than the government administration. *Lembaga adat* were believed by local people to be the primary decision makers in issues such as conflicts with logging companies (over 50 per cent of respondents), monitoring and management of natural resources (over 60 per cent of respondents), and development and protection of the National Park (over 50 per cent of respondents).

Adat Criteria for Natural Resource Management

Customary law — the part of *adat* concerned with sanctions — regulates access to and exploitation of land and forest products with regard to all forms of land and forest tenure. *Adat*, however, is neither fixed nor unchanging. As a social mechanism and judicial process, *adat* is transformed and adapted to new conditions in a constant evolution.

At annual meetings, which usually coincide with the harvest festival, members of the customary councils meet to discuss and update regulations and deliberate on social matters and natural resource management. Modifications to the regulations are often necessary as a result of changing circumstances, the negative effects of intensified harvesting pressure, or other changes in the natural environment and economic conditions.

Customary regulations specify modes of collection of forest products that tend to stress sustainability. Regulations emphasise not collecting more animals or forest products than needed, or harvesting in ways that would hamper their future reproduction or growth (for example, collection of hardwood resin is allowed throughout the entire village territory as long as trees are not cut down). In other instances, regulations may set temporal limits by determining how frequently a certain product may be harvested and for how long. With regard to rattan, for example, collection may occur only every two or three years. The period of collection is also limited to a two to three-week period. Modes of exploitation that employ chemicals and sophisticated technology which may have a damaging effect in the long term are outlawed.[1]

---

[1] For example, fish can only be caught using traditional tools like nets, rods, and fish traps, while the use of chemical poisons or electric shocks to catch fish is not permissible and offenders will be fined.

In recent deliberations of some customary councils, the hunting and trapping of animal species perceived as locally scarce was temporarily banned. These included the rhinoceros (a species that has supposedly been extinct in this area of Kalimantan for the last 40 years), clouded leopard, wild cattle (*Bos javanicus*), and the straw-headed bulbul (*Pycnonotus zeylanicus*). The latter, a common bird in the area of the National Park only a few years ago, has recently become a coveted item in wildlife trade and can fetch very high prices on regional and national markets.

Customary regulations commonly state that trees at the headwaters of rivers may not be cut. They also recommend that salt springs in the forest, which are common hunting grounds, may not be damaged by cutting the trees around the springs. While the first directive indicates a strategy of watershed protection and preservation of clean water supply for the community, the second one highlights the importance of protecting the habitat as a site of interrelated ecological and economic functions.

Based on these and other elements, it appears that communities are concerned with renewable supplies and the need to secure the future availability of natural resources for both consumption and commercial purposes. The regulations thus reflect what Western (1994: 500–1) calls a utilitarian value of conservation. Moreover, the emphasis on sustainability in the management of natural resources is a function of the current economic priorities based on the exploitation of forest products. However, if priorities change and the communities start to value the resources in their environment less, the interest in sustainable harvesting may cease.

Forest products with high use and market value, such as rattan, construction timber, fish, *gaharu* and other minor forest products, are commonly regulated but not to the same degree. For example, rattan — a diverse group of climbing palms — is a wild resource that is sometimes managed in locations where the resource is particularly abundant. Collection may be done only on a collective basis and upon deliberation of the community council. The harvesting is also limited to the older stems of the clump cut at a certain height from the ground, so that the rattan can grow back. On the other hand, *gaharu* — the resinous and fragrant heartwood produced by a fungus in trees of the *Aquilaria* genus — appears to be only slightly regulated. It is mandatory for *gaharu* collectors to report and pay a fee to the community council before going on a forest expedition, but collection may be done on an individual basis and at any time. Collectors are also expected to cut only infected trees.

It can be argued that differences reflect the products' natural characteristics and distinct population dynamics (Jessup and Peluso 1986), as well as local histories of control and exploitation. On the one hand, rattan tends to concentrate in certain forest areas and becomes a semi-managed resource under traditional tenure. Rattan is a rather predictable resource in terms of location and yield. More rules are therefore necessary to prevent over-exploitation of the resource and guarantee its sustainable use. On the other hand, *gaharu* trees are dispersed fairly evenly and at low densities over the territory. Regeneration from seeds takes a very long time, and trees can be found in a variety of habitats. Moreover, only one out of 30 trees may be infected. Consequently, *gaharu* is highly unpredictable, the output of collection activities is more uncertain, and the risk of depletion is therefore relatively low.

The history of exploitation of the two resources is also very different. Rattan is a forest resource that has been consistently used at the local level and commercially exploited on occasions (between 1910 and the 1960s, and again in the 1980s) depending on favourable market prices. By contrast, the high commercial value of *gaharu* and its large-scale exploitation are a relatively recent phenomenon (since the early 1990s) in the communities of the interior of Indonesian Borneo, and there is also no local use of *gaharu*.

The Management of National Parks in Indonesia

Official government regulations outline the legal and scientific framework for conservation management that is used and applied in all protected areas in Indonesia. These regulations are stipulated in documents like the *Government Regulations on Conservation and Protected Areas* (Regulation No. 68/1998) and the *Special Directive on the Management of National Parks* (attachment to Decree No. 129/Kpts/DJ-VI/1996), or the 1995 *Directive for Determining the Zonation of a National Park* (*Pedoman Penetapan Zonasi Taman Nasional*). Implementation is the responsibility of the Ministry of Forestry through the Directorate General of Forest Protection and Nature Conservation.

None of these documents explicitly deals with the rights of *adat* communities in conservation areas. The zonation system, however, comprises a 'traditional use zone' (*zona pemanfaatan tradisional*), where traditional activities and limited uses of plants and animals by local residents who are dependent on forest products are allowed (see also Regulation No. 68/1998 *Concerning Wildlife Reserves and Environmental Conservation Zones*). No animals protected by national law may be hunted, and only non-timber forest products may be harvested. One of the defining criteria for 'traditional' activities allowed in a National Park is the mandatory use of traditional tools like fishing rod and net, bow and arrow, or blowpipe and spear. Also, extraction or collection of forest resources should exclusively be for subsistence purposes or ceremonial *adat* needs. The Manual further indicates that only local people residing in the area are eligible for permits to use natural resources in the protected area, and the permits have to be granted by the park authorities.

Recent developments in forest policy include the ministerial decree (No. 677/Kpts-II/1998) on community forest concessions through the establishment of local cooperatives.[2] An internal draft instruction (Government of Indonesia 1999) discusses some possible new guidelines for use of natural resources in protected areas. Following the main directives of the ministerial decree, the draft proposes that the use of natural resources in protected areas be regulated as follows:

- exploitation of natural resources in ways and modes that are compatible with the main function of nature conservation;
- hunting activities in hunting parks with traditional methods such as dogs, arrows, spears or knives;
- harvest of only non-timber forest products (natural latex, birds' nests, traditional medicines, algae, honey, fruit, vegetables, edible roots or tubers, rattan), which means that no *gaharu* or timber may be cut nor minerals exploited;
- management of eco-tourism, natural resources, and hunting by local people organised in cooperatives in the use zones of the park, according to the specific functions of the protected area;
- management rights given to organised groups of local people or cooperatives for a definite period of time (30 years).

As with communities' forest concessions, local people are allowed to operate small businesses and manage natural resources in selected parts of the conservation area by forming joint enterprises or cooperatives. Permitted activities are those defined as 'traditional' and compatible with the function of protection of endemic flora and fauna in the park ecosystem. The government maintains full jurisdiction over the area through the park authorities. Park authorities also retain the right to monitor and evaluate activities, and to suspend the operating licence of a cooperative. This creates an odd legal situation whereby local people who, based on the authority of customary law, are the owners and managers of their customary land, would have to apply to the Minister of Forestry for a permit to operate businesses in 'their' land.

The internal draft contains no direct or explicit mention of *adat* or indigenous institutions, except with regard to the denotation of local people entitled to form enterprises:

> ... Indonesian citizens who were born and still live in and around the conservation area and possess the characteristics of a *komunitas* because of their social closeness, similarity of interests and means of livelihood

---

[2] This legislation has been suspended, and a revised version (Decree No. 31/Kpts-II/2001) has been approved.

that depend on the exploitation of natural resources, a common history, special bond to the land... (Government of Indonesia 1999: D-1).[3]

The issuing of the 1999 forestry law and the approval of the government regulations on *adat* forest will provide a strong legal mandate for enforcing changes to the management of conservation areas. New arrangements and regulations will have to accommodate the interests and rights of indigenous people as well as those of forest protection.

Recent Legislative Developments and the Status of *Adat*

A discussion of alternative management of conservation areas needs to take into consideration the 1999 *Forestry Law*, as several aspects of the law, particularly the aspects concerning customary rights and *adat* forest, will define future park management policy. Similarly, a definition of the role of local communities in the management of 'national' natural resources needs to be related to the recognition of *adat* rights under decentralisation and regional autonomy.

The *Forestry Law* of 1999

Law No. 41/1999 reasserts the principle that all forest land in Indonesia is controlled by the state for the prosperity of its people (Article 4), including customary forest or *hutan adat*, where management, not property, can be devolved to an *adat* community (Article 5). Special management rights over forest land can be granted to educational or research institutions, social and/or religious organisations, and/or indigenous communities or *masyarakat hukum adat*.

The law also contains a definition of *masyarakat hukum adat* (Article 67), which claims that the state acknowledges and accommodates local *adat* rights as long as they exist and are legitimate and they do not conflict with national interests. If, in future, *adat* communities should no longer exist, the right to manage the forest would be returned to the state. The recognition of an *adat* community as well as its abrogation will be established in regional regulations.

According to this section of the law, legitimate *adat* communities are those where:

- the community is still organised or recognises itself as one association under a common law (the Dutch *rechtsgemeenschap*);
- there is an active institution and officers;
- there is a clear territory controlled by *adat* (*wilayah hukum adat*);
- there is legal enforcement (and legal institutions/regulations);
- the community members still harvest forest products for their daily needs.

---

[3] *Komunitas* is a term often used in scholarly papers and official documents to denote *masyarakat adat* (traditional communities).

Legitimate *adat* communities also have the right to:

- use and exploit forest products for a living to meet their daily needs;
- manage the forest on the basis of existing customary law as long as it does not conflict with the national law.

Moreover, Article 34 of the *Forestry Law* states that the history of local communities and their institutions must be considered, as well as their record in management and conservation of the ecosystem. Although not directly related to the management of protected areas, the statement provides a strong mandate for the recognition and involvement of local institutions in the management of forests where such institutions exist.

A contentious issue remains to be resolved in that the legal existence of the *adat* community is contingent on its recognition and legitimation by the government. Similarly, the government would also decide whether *adat* rights would be abolished when *adat* institutions ceased to exist. The relevance of *adat* is thus subject to legal provisions outside the context of traditional law. Although there is an explicit recognition of *adat* claims over forest land, this is done within the framework or nomenclature of a state forest (Nugraha 2000: 3).

This situation has the potential to undermine the authority of *adat* and the sustainability of arrangements involving *adat*. Chris Bennett argues that:

> the key to a positive outcome is for *adat* or long established institutions to gain their legitimacy from below and from above, and allow the local community to decide which of its *adat* institutions should be formally recognised (personal communication, February 2001).

A draft government regulation on *adat* forest (Government of Indonesia 2000), which is currently being discussed, reinforces the basic principle of authority that *adat* forest is state forest. The draft specifies steps that need to be taken to recognise the existence of *adat* communities and establish the legitimacy of *adat* claims. It also clearly states (in Article 3) that *adat* communities that no longer exist cannot be re-established or resurrected. The Minister and regional government will form a research team comprising various experts in relevant disciplines who will determine the following:

- membership of the *adat* community;
- organisation and structure;
- boundaries of customary land;
- legal practices;
- management practices regarding forest products used in daily life and/or the cultural and religious relevance of *adat* forest;
- the history of the *adat* community.

The research methodology will be discussed and agreed between the Minister and the Indonesian Institute of Sciences.

As a cautionary note, potential shortcomings of this process must be indicated. For example, there might be a tendency to develop a standard methodological approach and impose it without due consideration of the local context. Moreover, the research process to establish the existence of an *adat* community and *adat* rights might take a very long time and entail high costs if it is to fulfill basic social science research standards, and ensure quality and reliability of results. There is a risk that short-term and superficial surveys by outsiders might be used to research the legitimacy of *adat* claims in order to cut costs and expedite the process. Moreover, there is no clear indication in the current draft of the government regulations whether existing documentation on *adat* communities and their claims to customary lands would be accepted by the government as valid. This would include evidence such as land-use and resource maps, customary regulations, and historical and cultural traditions. For example, in the Kayan Mentarang National Park area, the WWF project has already worked with the communities to compile thorough documentation on the existence of *masyarakat hukum adat* and the legitimacy of their claims over forest land by means of: long-term interdisciplinary research (see Eghenter and Sellato 1999, 2003); participatory community mapping activities (Sirait et al. 1994; Eghenter 2000b); and inventories of *adat* regulations and local institutions (Lawai 2001).

Decentralisation and the Management of National Parks

The law on decentralisation and regional autonomy (No. 22/1999 and No. 25/1999) concerns the transfer of political and financial powers from national or state level to sub-national or regional level. In this framework, reference to conservation and management of natural resources is minimal (Articles 7 and 10). The law states that the management of natural resources is transferred to regional (provincial and district) governments, but conservation policy and the authority over the management and development of protected areas remain the full responsibility of the Ministry of Forestry and Plantations.

The law contains some ambiguity with regard to the separation of jurisdiction between national and regional authorities in the management of natural resources. It also raises some questions concerning the level of decentralisation, whether at provincial or district level, for certain functions. According to Sembiring (2000), this ambiguity might create confusion and undermine the process of decentralisation unless it is improved in future drafts of the basic law or by further government regulations.

At a workshop in 1999, organised by the United States Agency for International Development and funded by the Natural Resource Management Project, several experts met to discuss what kind of management models would better suit the decentralisation scenario and guarantee more efficient management of National Parks. Saruan (1999) argued that management of National Parks in the new reality of decentralisation and regional autonomy would have to take into account the development plans of the regional or district government. In his view, these levels of government should be actively involved in setting up an efficient and transparent management system. Planning for the management of a National Park should follow a bottom-up approach and give priority to community-based models of conservation, where local conservation measures would be strengthened in the conservation area. Saruan further argued that the provincial office should be granted management autonomy while the central agency could act as a coordinating unit.

The integrity of a National Park in the future will not only depend on the effectiveness of conservation policies and application of existing laws, but also on compatible district legislation developed by the local parliament for the management of national natural resources which are located in its territory. The drafting of district regulations provides a good opportunity for conservation managers to work together with the local parliament on specific mechanisms outlining the role of the regional government in managing 'national' protected areas and for exploring collaborative institutional arrangements that would include the district government as part of the managing unit.

The *Masyarakat Adat* Management Model

The preceding discussion establishes that there is a missing link between *adat* regulations and national regulations, between the legal framework of customary law and that of national law, with regard to the protection and management of conservation areas. However, it also reveals the potential points of convergence between the two perspectives and indicates the need for new models and legal avenues to create effective and equitable 'localised' management of protected areas. The proposed model would be 'localised' in that it would take into consideration the aspirations of local people for improving their welfare and taking part in the management of a protected area. It would integrate existing local *adat* institutions and regulations on sustainable use of natural resources as part of regional conservation efforts. There is not necessarily a contradiction between the efforts to conserve a forest area and local forms of exploitation. This is particularly true for areas like the Kayan Mentarang conservation area, with a history of sustainable use of natural resources (under stable conditions), low population density, little agricultural pioneering or illegal hunting, and where risks of over-exploitation are more likely to originate from outside the area.

The new forestry law explicitly states the criteria by which the government can recognise the legal existence of *adat* communities. These criteria include: the relative sustainability in the use of natural resources; the presence of strong social cohesion and traditional institutions; high reliance of communities on the exploitation of natural resources; and a tradition of conservation measures. In the case of a conservation area that is claimed by *adat* communities, like the Kayan Mentarang National Park, an additional criterion for the acknowledgment of the legitimacy of *adat* claims would need to be considered. This criterion would be the level of dependence of the local people on the objectives and success of the project. The communities living in and around the Kayan Mentarang area are not only *adat* communities with functioning traditional institutions, customary territories, and a long history of occupation and use of the area. They are also economically dependent on the extraction of valuable forest resources from the park area and the conservation of its forest status. For example, the support for the protected area is highest among those communities who are most economically dependent on forest resources, so long as local communities are allowed to continue sustainable extraction. There is a strong correlation between the economic value of the forest (in the form of non-timber forest products) and support for the establishment of a National Park (which is the main objective of the project).

Some preliminary recommendations can thus be made in support of localised conservation agendas:

1. Secure official recognition of *adat* land and building the capacity of customary councils in their role as managers of the conservation area.
2. Preserve locally developed regulations on the use of forest products that guarantee sustainability, including suggestions and criteria for animal-population management. This strategy is likely to increase the chances of compliance among local people.
3. Accept *de facto* core zones as those areas which are too far from settlements and are not exploited by local people, but which would maintain important ecological functions in the conservation area.
4. Recognise that definitive and precise entitlements are probably more useful for communicating boundaries to outsiders and discriminating between *adat* and non-*adat* people (users and outsiders) than they are as criteria for management of the conservation area.
5. Create an inter-*adat* institution or forum that coordinates management activities and addresses environmental concerns that often transgress the local boundaries of customary lands.
6. Recognise that National Parks established in territories claimed by indigenous people are best managed and protected as indigenous or *adat* forest.

It is recommended that customary councils or *lembaga adat* be recognised as managers of the *hutan adat,* which is part of the National Park area. *Lembaga adat* are active and established institutions with a strong basis of tradition and legitimacy at local level. They, and the communities they represent, 'have the same interest in securing access and use of natural resources and the ecosystem', which is one of the criteria discussed in the internal draft of guidelines to regulate community forest concessions in conservation areas. Moreover, the tradition of the customary councils in the area of the National Park indicates that they possess a strong commitment to protecting the environment and practising sustainable use. They also have knowledge and experience in the management of natural resources, which are additional requirements mentioned in the draft (Government of Indonesia 1999).

It is important to develop and enforce an *adat*-based management of the park by training and supporting local institutions. The opportunity for capacity building would strengthen local management and legitimise the role of local people, not just as simple users but also as managers of (their) natural resources in the park area. The process would take time and it would have to be adjusted to suit the ability and time constraints of the communities.

Developing an *adat*-based management of the park would also indirectly strengthen and reinforce a new social role for the customary councils at a time of extreme challenges and difficulties for *adat*. The example of the exploitation of *gaharu* in Apo Kayan is in many ways typical of the exploitation of natural resources in the interior of Kalimantan. Since the early 1990s, outside collectors, sponsored by Chinese and Arab traders based in the towns of the lowlands, have come in increasing numbers in search of *gaharu* and gallstones. Their mode of exploitation is drastically different from local practices. Being outsiders and belonging to different ethnic groups, they do not always acknowledge or respect local *adat* regulations and rights. They tend to cut indiscriminately infected and non-infected trees, and use chemicals and other means to poison salt springs where langur monkeys come to drink. They also spend extended periods of time in the forest where they establish semi-permanent camps. This mode of exploitation has increased the chances of over-exploitation of natural resources and has also exposed the limits of the local *adat* authority. For example, the customary councils often deliberate on the need to prevent outside collectors from accessing their land, and the options of confiscating the collectors' supplies and belongings. They denounce the situation but sometimes lack the necessary legal authority and internal consensus to impose their will. Enforcement of, and compliance with, regulations is an index of the strength of local *adat*. When traditional authority begins to lose its prestige and is eroded by competing normative systems, the effectiveness of the local management system is also inevitably weakened and may collapse (Eghenter 2005).

The WWF Kayan Mentarang Project has compiled and proposed a preliminary conservation agreement between the communities and the Directorate General of Forest Protection and Nature Conservation for the management of the park based on local *adat* regulations. Its conceptual and practical justification draws upon the considerations of the local economic and social situation: the need to recognise and secure the exclusive and permanent usufruct rights of communities in the area of the park; the relevance of customary regulations where these stress conservation and sustainable use; the importance of allowing local enforcement and the imposition of customary fines for most infractions; the introduction of tools such as quotas and seasonal harvesting, or other measures of animal-population management, when and if conditions require. The conservation agreement has already been discussed in local meetings with the communities and the feedback was positive. The communities felt that their concerns were being addressed and that they could support plans for a National Park based on recognition of their *adat* claims and customary law (Eghenter 1999).

Participants in the 1999 workshop suggested that a management plan with a clear zonation system and division into core zone, wilderness zone, and traditional-use zone would help acknowledge, integrate, and accommodate the conservation functions of the protected area and the aspirations of local people. However, this recommendation alone might not be enough to achieve these objectives. Zonation should be established in ways that best suit local conditions. In this regard, not all kinds of zones might be appropriate in a given protected area, but only those most relevant for maintaining functions of biodiversity protection and securing the economic needs of local people. For example, in the Kayan Mentarang National Park, the entire area is claimed by *adat* and most of it has been exploited in the course of history. In these circumstances, the establishment of a large traditional use zone or '*adat* use zone' might represent a priority, while a core zone would only be determined following an assessment of local land-use patterns and trends, and on the basis of wide local consensus.

Another important consideration is that a zonation system imposes a sort of permanent micro-partition within the conservation area according to ecological, biological, research, and other functional criteria. This approach, unless it is the result of a consultation process and linked to local standards of land use, can raise suspicion among local people. For example, during participatory planning meetings for the Kayan Mentarang National Park, community representatives objected to the proposal to divide their territory into different zones, each with its own separate set of regulations and prohibitions. Moreover, they indirectly questioned the meaning of a permanent division into zones by asking, 'once a zone has been established, can we change it?' or, 'can we access a core zone once we have exhausted all resources in the other zones?'

In the recommended 'localised' management model, the day-to-day management of the park would be the full responsibility of the inter-*adat* institution — a coordinating institution formed by elected members from each of the customary councils in and around the conservation area. The creation of this institution would guarantee easier coordination between the various *adat* units and better overall management and use of the entire area of the park. In the case of the Kayan Mentarang National Park, this institution would be the Forum Musyawarah Masyarakat Adat that was formally established in 2000.

In regard to the ideal role of central and regional governments in the management of National Parks, Yusuf (2000: 49) suggests that the central government only act to facilitate, advise, and provide guidelines. In 2000, the forum members made a recommendation to the Directorate General of Forest Protection and Nature Conservation with regard to recognising their role as managers of the park. Subsequently, it was proposed that a collaborative form of management be established. The proposed Dewan Penentu Kebijakan (Policy Board) was to include representatives and conservation experts of the ministry, representatives of the forum or the indigenous people of the park area, and representatives of the local government. In April 2002, the policy board was formally recognised by a ministerial decree for the collaborative management of the Kayan Mentarang National Park. The operating principles of the board emphasise the importance of coordination, competence, shared responsibilities, and equal partnership among all stakeholders.

The *adat*-based management model of conservation areas could have important social, economic, and ecological advantages. With the involvement and acknowledgment of local people as managers of the park, there would be reduced initial costs for activities like building, monitoring, boundary marking, and law enforcement. The implementation of this kind of management would entail a smaller opportunity cost for local people and avoid significant social costs. Local residents would be granted exclusive rights to use the forest sustainably and sell forest products. The legitimation of *adat* would guarantee a degree of tenure security to local communities. Moreover, their presence on the management board as equal partners could enhance their sense of responsibility and accountability in the management of the forest.

Postscript (May 2004)

This chapter was originally completed in 2001. Although the discussion and challenges regarding the relationship between customary law and national law in the management of National Park areas remain valid, there have been some interesting developments in the meantime. With regard to the management of conservation areas, the most interesting aspect is the forthcoming Ministry of Forestry directive on collaboration in management activities in protected areas (*Pelaksanaan Kolaborasi Kegiatan Pengelolaan Kawasan Suaka Alam dan Kawasan Pelestarian Alam*). The principle of collaboration and 'stakeholder theory' would thus be established as conditions for more effective management of conservation areas in Indonesia. As mentioned in the section on basic principles for collaboration, it is suggested that the form of collaboration be adapted to the social, cultural, and ecological conditions of protected areas. Interestingly, this aspect seems to further underline the need to 'localise' park management.

References

Casimir, M., 1992. 'The Dimensions of Territoriality: An Introduction.' In M. Casimir and A. Rao (eds), *Mobility and Territoriality: Social and Spatial Boundaries among Foragers, Fishers, Pastoralists, and Peripatetics.* New York and Oxford: Berg.

Chartier, D. and B. Sellato, 1998. 'La Prise en Compte des Pratiques et des Usages Autochtones: Realité Efficiente ou Construction Occidentale a Visée Neoliberale [Accounting for Indigenous Practices and Customs: Effective Reality or Western Construction through a Neoliberal Lens].' Paper presented at a conference on 'Social Dynamics and Environment', Bordeaux.

Eghenter, C., 1999. 'Planning for Community-based Management of Conservation Areas: Indigenous Forest Management and Conservation of Biodiversity in the Kayan Mentarang National Park, East Kalimantan, Indonesia.' Paper presented at a conference on 'Displacement, Forced Settlement and Conservation', Oxford, 9–11 September.

————, 2000a. 'What is *Tana Ulen* Good For? Considerations on Indigenous Forest Management, Conservation, and Research in the Interior of Indonesian Borneo.' *Human Ecology* 28: 331–357.

————, 2000b. 'Mapping Peoples' Forests: The Role of Mapping in Planning Community-based Management of Conservation Areas in Indonesia.' Washington (DC): Biodiversity Support Program.

————, 2005. 'Histories of Conservation or Exploitation? Case Studies from the Interior of Indonesian Borneo.' In R.L. Wadley (ed.), *Histories of the Borneo Environment: Economic, Political, and Social Dimensions of Change and Continuity.* Leiden: KITLV.

———— and B. Sellato (eds), 1999. *Kebudayaan dan Pelestarian Alam: Penelitian Interdisipliner di Pedalaman Kalimantan [Culture and Environmental Sustainablility: Interdisciplinary Research in Interior Kalimantan].* Jakarta: Ford Foundation and WWF Indonesia.

————, 2003. *Social Science Research and Conservation Management in the Interior of Borneo: Unraveling Past and Present Interactions of People and Forests.* Bogor: Centre for International Forestry Research (CIFOR), Ford Foundation, UNESCO, and WWF Indonesia.

Government of Indonesia, 1999. 'Petunjuk Tehnis Pelaksanaan Hutan Kemasyarakatan di Kawasan Pelestarian Alam [Technical Instruction for the Execution of Community Forest in Environmental Conservation

Zones].' Bogor: Departemen Kehutanan, Direktorat Jenderal Perlindungan Hutan dan Pelestarian Alam [Department of Forestry, Directorate General for Forest Protection and Environmental Conservation].

————, 2000. 'Rancangan Peraturan Pemerintah Republik Indonesia, Tahun 2000, Tentang Hutan Adat [Proposed Government Regulation of the Republic of Indonesia, in the Year 2000, Concerning Traditional Forest].'

Jessup, T. and N. Peluso, 1986. 'Minor Forest Products as Common Property Resources in East Kalimantan, Indonesia.' In *Proceedings of the Conference on Common Property Resource Management*, 21–26 April 1985. Washington (DC): National Academy Press.

Lawai, L., 2001. 'Peranan Lembaga Masyarakat dalam Pengelolaan Taman Nasional. Analisa Hasil Inventarisasi Partisipatif Lembaga Lokal di Kawasan Kayan Mentarang [The Role of People's Organisations in Managing National Parks. An Analysis of Findings of an Inventory of Participatory Local Organisations in the area of Kayan Mentarang].' Unpublished report to WWF Indonesia.

Nugraha, A., 2000. 'Implementasi Konsep Pengelolaan Hutan Berbasis Masyarakat di Era Otonomi Daerah: Suatu Kajian Dalam Perspektif Pengusaha Hutan [Implementing the Concept of Community-Based Forest Management in the Era of Regional Autonomy: A Contribution from the Perspective of a Forest Entrepreneur].' Paper presented at a workshop on 'People's Initiatives in Managing the Natural Resources of East Kalimantan', Samarinda, 22–23 August.

Saruan, J., 1999. 'Dukungan Pemda Mengintegrasikan Pembangunan Wilayah dengan Pengelolaan Taman Nasional dalam Antisipasi Pelakasanaan Otonomi Daerah [Endorsement of Regional Government for Integrating the Development of the Region with the Management of National Parks in Anticipation of the Implementation of Regional Autonomy].' Paper presented at a workshop on 'Managing National Parks in the Eastern Region of Indonesia', Manado, 24–27 August.

Sembiring, S., 2000. 'Desentralisasi-Desentralisasi Pengelolaan Sumberdaya Alam: Apa Yang Diperoleh Rakyat? [Decentralised Management of Natural Resources: What Do the People Get Out of It?]' In S. Sembiring et al. (eds), op. cit.

————, N. Makinuddin, Phantom, E. Marbyanto and S. Raharjo (eds), 2000. *Menjadi Tuan di Tanah Sendiri. Menuju Desentralisasi Pengelolaan Sumberdaya Alam Kalimantan Timur [Taking Control Over Your Own Land: Aiming for Decentralised Management of the Natural Resources of East Kalimantan]*. Samarinda: WWF-KAN, NRMP, APKSA dan PEMDA Propinsi Kaltim.

Sirait, M., S. Prasodjo, N. Podger, A. Flavelle and J. Fox, 1994. 'Mapping Customary Land in East Kalimantan, Indonesia: A Tool for Forest Management.' *Ambio* 23: 411–417.

Western, D., 1994. 'Linking Conservation and Community Aspirations, in Natural Connections.' In D. Western and M. Wright (eds), *Natural Connections*. Washington (DC): Island Press.

WWF (Worldwide Fund for Nature), 1996. 'Indigenous Peoples and Conservation: WWF Statement of Principles.' Gland: WWF International.

————, 1998. 'Principles and Guidelines on People and Forests.' Gland: WWF International, People and Conservation Unit (draft proposed to the WWF/World Bank Forest Alliance).

Yusuf, A.W., 2000. 'Sistem Desentralisasi dalam Pengelolaan Sumberdaya Alam dan Lingkungan Hidup Menurut UU No. 22/1999 Tentang PEMDA [The Decentralized System in Managing Natural Resources and the Environment through Regulation No. 22/99 According to PEMDA].' In S. Sembiring et al. (eds), op. cit.

Chapter Nine

The Potential for Coexistence between Shifting Cultivation and Commercial Logging in Sarawak[1]

Mogens Pedersen, Ole Mertz and Gregers Hummelmose

Introduction

Shifting cultivators and logging companies have traditionally been considered to be in conflict with each other as they are using the same resources for different purposes. Shifting cultivation is usually a subsistence-oriented agricultural system clearing forested areas for fields with annual crops and leaving these areas fallow for varying periods. Logging is mostly a purely commercial activity using the largest valuable trees, while the logged-over areas are either left for regrowth and re-logged after a number of years or else clear-cut for development of forest plantations or industrial crops such as oil palm or rubber.

Criticism of both systems has come from different sides. Shifting cultivation has mostly been accused of being wasteful of natural resources, having low productivity and maintaining people in a vicious circle of poverty (FAO Staff 1957; Lau 1979; Watson 1989; Rasul and Thapa 2003), and negative views of this farming system persist in many government circles in countries where shifting cultivation still occupies relatively large areas (Fox 2000). Conversely, logging activities have been under severe criticism by green organisations as well as from various academic writers — a criticism not focused exclusively on the impacts on the physical and biological environment, but also on the jeopardised livelihoods and land rights of communities in areas affected by logging (Hong 1987; Colchester 1993; Jomo 1994).

Colchester (1993), for example, argues that the system of Native Customary Rights Land in Sarawak leaves the natives without clear rights to what they perceive as their land, and their rights are not adequately acknowledged when

[1] This study was funded by the Danish University Consortium on Sustainable Land Use and Natural Resource Management, under the Danish Cooperation for Environment and Development Program of the Ministry of Energy and Environment. The authors would like to thank the Sarawak State Planning Unit and the Forestry Department of Sarawak for supporting the project and offering invaluable assistance during fieldwork. We would also like to express our gratitude to the people living in Rumah Chili and Rumah Agau whose hospitality and openness made this study possible and worthwhile, and to the entire staff at the Sekawi Logging Camp. Finally, we would like to thank the organisers of the conference on Resource Tenure, Forest Management and Conflict Resolution. Perspectives from Borneo and New Guinea, Canberra, 9–11 April 2001, for valuable comments on the chapter.

concessions are given.[2] This has led to land use conflicts, and as individual communities often lack the power to influence decisions made on land use issues, partial alienation of land has been the result. Moreover, Colchester (1993) argues that Iban community leaders have been working more for their own benefit than for their community as such, and therefore have not adequately assisted local populations in land disputes. Local resistance towards logging has mostly been reduced to blockades of logging roads, which have been rapidly reopened by the police. Colchester shows how disputes over land date back to pre-colonial patterns of state control over forest resources, and how difficult it can be for local people to resist development plans involving their land.

The other extreme is represented by Lau (1979), who declared that shifting cultivators pose a threat to state interests as they cause 'wanton destruction' of valuable timber resources and are responsible for soil erosion, pollution and siltation of waterways, pollution of the air, river flooding, and the loss of valuable genetic resources and habitats for wildlife. His main concern was related to the rapid destruction of primary forest, but the calculations behind the data have been questioned by Hurst (1990), who sees them as overestimates, and by Hong (1987), who states that the calculations ignore the fact that shifting cultivators frequently prefer secondary forest for cultivation. However, the negative attitudes towards shifting cultivation within the governmental structures of Sarawak are not surprising, given the large revenues from the export of timber and the number of people employed in the forestry sector. Moreover, politics and logging have always been inextricably linked to each other (King 1993; Ross 2001) and, as stated by King (1993: 242):

> The arrangements with Chinese entrepreneurs are an important means to cement cross-party alliances between Bumiputra and Chinese political leaders; these alliances are essential in the context of Sarawak's political system.

Other authors, such as Dauvergne (1997), widened the frame of explanation to include international perspectives in the analysis of forest exploitation in Southeast Asia. Dauvergne claimed that large Japanese conglomerates control the logging operators in Sabah and Sarawak through favourable credit arrangements and thereby increase their logging rates without promoting sustainable forest management.

The opposing views on shifting cultivation and logging still persist in rather uncompromising forms, although a number of studies have tried to soften the conflict by a more balanced analysis of the systems (King 1993; Potter 1993). However, the question of whether peaceful coexistence between these two land

---

[2] For further description of the *Sarawak Land Code,* see Cramb and Wills (1990) and Cleary and Eaton (1992).

use systems can be achieved seems to have been neglected. The objective of this chapter is to investigate interactions between natural resource managers in an area in Sarawak where Iban shifting cultivators live side by side with a large logging concession. We analyse the socio-economic and perceived ecological impact of the logging operation on the Iban communities as well as the effects of Iban shifting cultivation on logging. The potential for improved coexistence between these systems is discussed on the assumption that both shifting cultivation and logging are likely to continue in the future, and it is therefore counterproductive to focus only on the negative effects and interactions rather than on the opportunities for harmonising the two systems.

Study Area and Methodology

The present study focuses on the Model Forest Management Area (MFMA) and surrounding lands in the Muput area southwest of Bintulu in the Bintulu Division, Sarawak (see Figure 9.1). The MFMA was jointly established by the Sarawak Forest Department and the International Tropical Timber Organisation in 1996 following the report of a mission to Sarawak in 1989/90 which suggested the introduction of more-sustainable forest management (ITTO 1990). Physical and socio-economic studies of local communities in the area were carried out before its establishment (Sidu 1995). The MFMA supports training, demonstration and research on sustainable hill-forest management, and the sustainable logging methods practised include improved planning of roads and skid trails, and extra care for the residual stand when felling and removing trees. The logging operation is run by several companies, the closest to the two communities studied in this chapter being Zedtee, which has its local headquarters at Sekawi Camp (see Figure 9.1).

Some 40 Iban longhouses are located in and around the MFMA, and this study takes as its point of departure two of these — Rumah Agau and Rumah Chili (see Figure 9.1). These Iban settlements have been present in the area since the early 20th century, while logging has been carried out since 1976. The longhouse inhabitants hold their land under Native Customary Rights, whereas the area within the MFMA is designated as part of the Permanent Forest Estate (Sidu 1995).

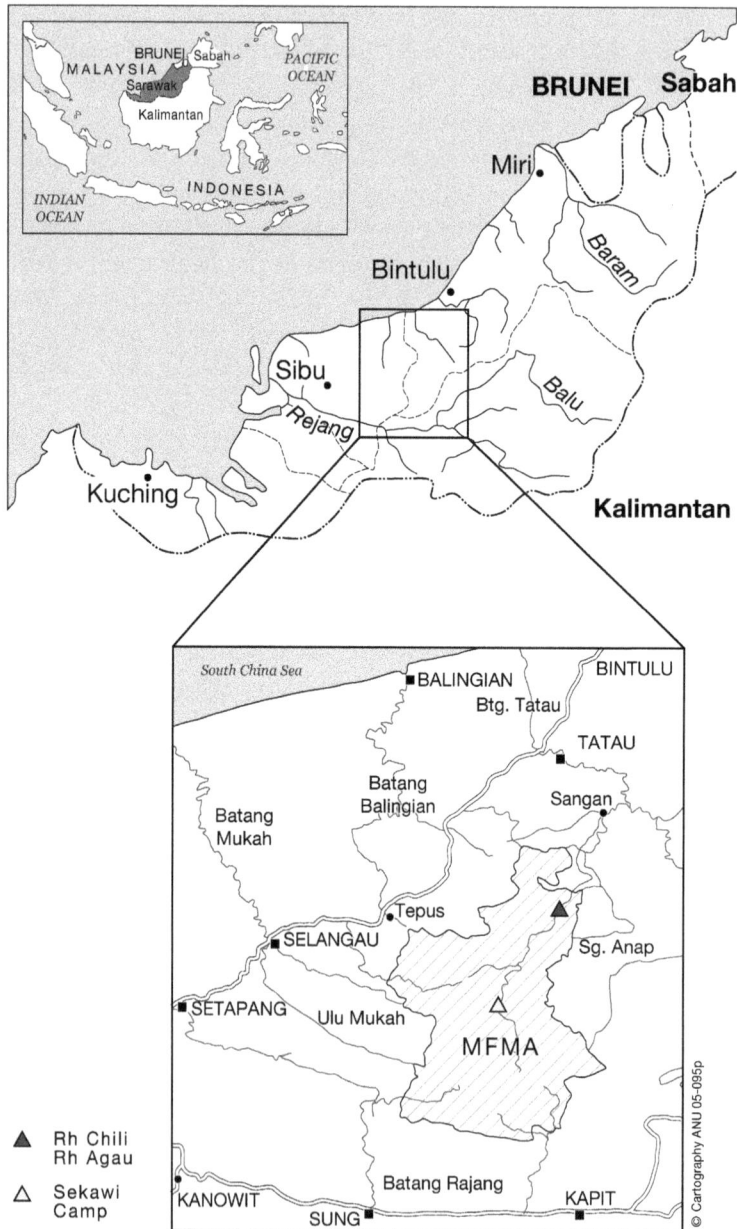

Figure 9.1. Location of Iban communities and the Sekawi logging camp in the Model Forest-Management Area

Data were collected between June and September 1999. A structured questionnaire survey was carried out covering all permanent-resident households in the longhouses — 17 households in Rumah Agau and 14 households in Rumah Chili. The topics addressed were household composition, wealth, farming practices and off-farm activities. Semi-structured in-depth interviews focusing on farming strategies and perceptions of logging were carried out with five households in Rumah Agau and four in Rumah Chili. Finally, four focus group interviews were carried out in each community covering the following topics: forest products, off-farm labour, land tenure, and perceptions of logging.

Thirty semi-structured interviews were conducted among employees in the logging operation. These interviews included management staff as well as labourers, and were carried out within the logging camp and in the blocks logged at the time of study. Finally, structured interviews with government officers at different levels in the Forest Department were carried out. Follow-up interviews were conducted with several of these officers when needed.

Results and Discussion

The main finding of this study is that conflicts over land between shifting cultivation and logging in the area were insignificant. The logging operator was viewed positively by most of the Iban interviewed because of job opportunities and some infrastructural assistance, even though the operation was accused of having a negative impact on hunting, fishing and the gathering of wild products. Similarly, the logging operator did not view the presence of the communities as a serious obstacle to the logging operation. These findings are explained in more detail below.

Changes in Natural Resource Management within the Study Area

Shifting cultivation in the study area operates almost without the use of primary forest. First, both longhouses have been situated in the area since the early 1900s and can be considered sedentary. Furthermore, as a result of out-migration, the population size has not changed much in the 20 years preceding the study. This means that the habit of pioneering shifting cultivation in search of primary forest is no longer needed, as native customary rights to a sufficient amount of land have already been established through the clearing of primary forest in the past. Second, the extra labour input required for felling primary forest is not compensated by a proportionate increase in yields. Third, as seen elsewhere in Sarawak (Mertz and Christensen 1997), the Iban increasingly prefer to cultivate closer to the longhouse where primary forests are no longer present (see Figure 9.2). This may be caused by the lack of labour in households dominated by elderly people, which is an indirect effect of the improvement of roads and hence the accessibility of the area. Whereas, in other parts of the world, the opening up of new forest lands is frequently accompanied by an influx of new settlers or resettled persons (Myers 1992; Whitmore 1998), in this case it seems that the movement is reversed and has led to a stabilisation or even reduction in the number of people living *de facto* in the communities. Good off-farm job opportunities in Malaysia are the main reason for this development.[3]

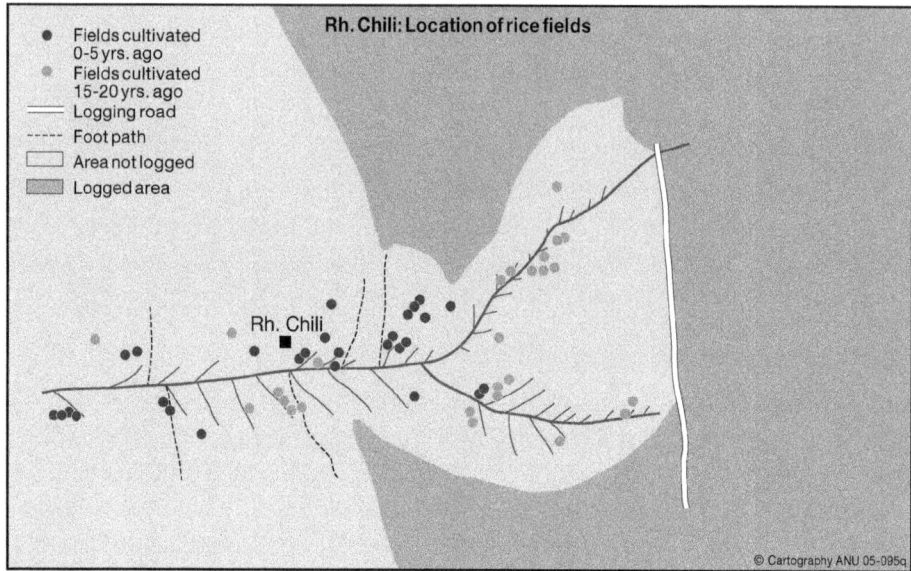

Figure 9.2. Location of cultivated rice fields in the Rumah Chili Area, 1974–79 and 1994–99

Logging was carried out prior to the establishment of the MFMA. Several areas close to both longhouses were logged from 1976 until 1996. In 1974 all longhouses in the area were asked to decide whether they would accept logging in their area or not. Meetings were held with the heads of longhouses and the *penghulus* (heads of several communities within one river catchment), who decided that, if compensated by the logging operators, they would accept logging in the area. Whether this decision was supported by the general Iban public in the area is not known.

Different logging operators initiated operations and some areas have been logged more than once. The compensation agreed upon included, among other things, a monthly payment to the longhouse head, a yearly payment to each household in the longhouse, and the landowners affected by road construction were paid compensation of one Ringgit per metre of road constructed on their land. In addition to this, some longhouses still receive diesel fuel for their generators and (to a minor extent) building materials from the operators.

Some 300 000 m^3 of logs are now felled annually in the 162 500 ha that make up the MFMA. Selective logging, as carried out in the MFMA, is scheduled to allow re-entry after 25 years of regeneration. Since the operation was initiated

---

[3] Similar trends in other parts of Southeast Asia have been described by Rigg (2001) and Breman and Wiradi (2002).

in 1976, most of the primary forest in the MFMA has already been logged, and the companies now focus on re-entering previously logged blocks rather than expanding into new areas. The logging operation is run from four main logging camps; the headquarters of the operation is the Sekawi camp located in the centre of the MFMA (see Figure 9.1). More than 600 people are employed in the operation, of whom approximately two thirds are Iban. The Iban are almost entirely engaged as chainsaw or bulldozer operators and surveyors. Only about half of the Iban workforce can be considered as locals whose homes are within a half day's travel of the main camp. Very few Iban have camp-based jobs such as those of managers, mechanics and shopkeepers. These jobs are generally more desirable as they are less hazardous and have fixed salaries with a retirement scheme, but jobs in the field-based logging operation, which are paid by the hour, can yield high wages if working hours include overtime.

The effects of low-impact logging as carried out in the MFMA have previously been described by Marn and Jonkers (1982). They show that low-impact logging can be carried out without jeopardising the economy of the operation, even while reducing damage to the residual stand. The number of trees damaged during the logging operation is still generally high, but there are different views on the extent of the problem. Kartawinata et al. (1981) and Whitmore (1998) estimate that 35–40 per cent of the residual stand is damaged during the extraction of 10 per cent of the trees. Contrary to this, Brown and Press (1992) claim that more than 70 per cent of the residual stand is damaged or destroyed by this level of extraction. These variations usually reflect the different ecological conditions and management decisions under which logging is carried out, but may in certain cases also represent different agendas and opinions on tropical rainforest utilisation.

All land within the MFMA is designated as part of the Permanent Forest Estate (Sidu 1995), despite the fact that several plots of land within the MFMA fulfil the conditions for being Native Customary Land. This has not as yet led to any conflicts. The logging operators are usually uninterested in areas used for shifting cultivation as they do not contain sufficient numbers of large, valuable trees to be profitable for logging, and regeneration of cleared plots to climax forest would take at least 60–80 years.

However, should the communities decide to claim land with logged-over forest as Native Customary Land and practise shifting cultivation in these areas, conflicts would be likely to arise because the logging operator expects to re-enter these areas after a period of 25 years. Although the extent of this practice is not known, and during the time of study no fields were located in logged-over forest, this issue is a major concern for the logging operator. The two longhouse communities had diverging opinions on the suitability of such areas. People in Rumah Agau perceived logged forest as unsuitable for cultivation due to the disturbance of the soil caused by the heavy machinery. On the other hand, people in Rumah Chili thought areas previously logged would be very suitable for cultivation, especially because clearing of vegetation had become easier due to the presence of smaller trees. Another potential point of conflict would be the establishment of forest plantations within the MFMA, including areas previously used for shifting cultivation, as this would permanently alienate land to which local communities may have claims for customary rights. However, such plans have not yet been initiated.

The Iban communities in the study area are still highly dependent on access to natural resources extracted from the forest and rivers, and the possibility for successful hunting and fishing has diminished since the logging companies entered the area. Siltation of rivers is unavoidable when logging operations are carried out (Douglas et al. 1993), and, according to local people, fish populations have declined as a consequence. Increased human activity, whether brought on directly or indirectly by the logging operations, can have the same effect on the number of hunted animals as these flee to areas less affected by logging. The change in the number of animals and fish was described by Aiken and Leigh (1992) as the result of several factors, including habitat change, hunting by timber company workers, greater availability of ammunition, improved access to forests along logging roads, high turbidity levels, siltation, and pollution from diesel oil in rivers. Rumah Chili people notably complained of decreasing food supplies from the forest and rivers and increasing reliance on purchased products.

Changes in the Socio-Economic Conditions of the Iban Communities

Declining availability of forest products has partly been offset by other activities, thereby diminishing the dependence on natural resources. The improved possibilities for off-farm employment have enticed many younger Iban from Rumah Agau and Rumah Chili to engage in jobs outside the longhouse community. Some people find work in towns or even overseas for shorter or longer periods, but most jobs are found in the timber industry. These jobs require little education, and those within the actual logging operation, such as chainsaw operation, require skills already held by many Iban men.

The ongoing logging operation has led to the construction of a main logging road running through the study area. This has improved market access considerably as the travel time to the nearest towns has been shortened. Furthermore, the logging camp itself is an important market place, and various products such as durian, vegetables and fish are sold along the logging road. Another effect of the easier market access is an increase in cash-crop production, mainly rubber and black pepper. Sidu (1995) found that cash crops were cultivated mostly by the communities with the best market access.

Cash-crop schemes are mainly introduced by the Department of Agriculture, and although the communities express an interest in these schemes, they have only been adopted by a few families. Cash crops tend to be labour-intensive, and their prices fluctuate, making them unreliable as a stable source of income (Cramb 1988; Wadley and Mertz 2005). Extended periods of low world market prices might explain why black pepper gardens in both communities sometimes lack maintenance or are even abandoned, and why potentially productive rubber trees are being felled and the land converted to hill rice fields.

A more recent, but economically important, activity for the communities is the substantial extraction of logs of the illipe nut tree (*Shorea macrophylla*), which has been made possible by an entrepreneurial middleman at the mouth of the Muput River. This activity was carried out by the majority of households in the two longhouses. For some households the selling of logs was the most important income-generating activity. The more than 500 tons of logs sold during 1999 must be expected to exceed the future amount available for extraction.

Interactions Between Actors and the Potential for Greater Coexistence

Many aspects of the coexistence between the Iban population and the logging operator seem to work well. One initiative improving the coexistence could be directed to finding remedies for the decline in availability of forest and river products. Construction of fish ponds has already been carried out with machinery from the logging operation, and an expansion of this activity seems obvious as it substitutes for the decline of fish in the rivers and may provide income-earning opportunities. More areas close to the logging road could be converted into fishponds with a minimum of effort if machinery from the logging operation were used more frequently. In line with this, a focus could be directed towards animal husbandry, thereby partly substituting the decline in the presence of wild animals. The Department of Agriculture has supported this activity through the Animal Husbandry Improvement Scheme, but only to a limited extent. The logging camp itself could be a potential market for such products.

The remote location of the communities means that off-farm employment normally implies moving away from the longhouse and only returning for a few days each month. The logging operator in the study area employs many people from the longhouses situated within or close to the MFMA, but the majority of the employees are from other parts of Sarawak, and a few even from Kalimantan. Employing people from areas far away is part of the policy of the logging operator as these people are perceived as a more reliable workforce. Employees are required to stay in the logging camp permanently and only return to their community for for or five days each month, but local employees tend to return to their home more often without approval from the camp manager. Although the Iban have a tradition of male labour migration (*bejalai*), and do not consider it a major obstacle to have to travel to get a job (Kedit 1993; Wadley 1997), the local population prefers to engage in jobs close to the longhouse. Furthermore, if the local workforce were able to return from the logging camp to their home once a week instead of once a month, it would be easier to fulfil obligations within the longhouse. The present routine seems to have been made with the intent to allow workers living far from the MFMA to return to their homes. Changing the employment policies and working routines could probably facilitate a higher proportion of local employees in the logging operation and further improve the coexistence between the local population and the logging operator.

As logging is a widespread practice throughout Sarawak and is probably bound to continue in the future, it is important to emphasise the positive impacts of concessions with long time spans (ITTO 1990). Long-term concessions would create an incentive for the concessionaire and logging operator to adopt low-impact logging procedures, thereby allowing for a second entry. At the same time, knowing that the operation will continue in the area for a long time should encourage both the local population and the logging operator to establish good relations and develop the potential for mutually beneficial coexistence.

Conclusion

Based on the interviews with two communities and staff in the logging camp, it can be concluded that coexistence between shifting cultivators and loggers in the study area is relatively smooth. At worst the longhouse inhabitants are indifferent to the logging operation, but most informants are satisfied with the logging activities, which have facilitated an improvement of the standard of living. At the same time the logging operator is indifferent towards shifting cultivation as long as it is carried out without the use of primary and logged-over forest within the MFMA.

The conflict scenarios presented by Lau (1979) and Colchester (1993) seem not to be applicable to the study area, and even though the MFMA is a trial area for sustainable forest management and as such could be considered an unrepresentative showcase, this study demonstrates the potential for mutually beneficial coexistence between actors traditionally considered to be in conflict with each other. It is by no means impossible that similar arrangements could be secured in other logging areas, particularly where the purpose of the logging operation is to maintain the Permanent Forest Estate for long-term timber production. In State Land Areas designated for conversion to plantations, the situation may be different, as communities are likely to be subjected to more pressure to engage in long-term leases of their customary land to oil palm plantations in joint venture arrangements (Majid Cooke 2002; Ngidang 2002).

References

Aiken, S.R. and C.H. Leigh, 1992. *Vanishing Rain Forests: The Ecological Transition in Malaysia.* Oxford: Clarendon Press.

Breman, J. and G. Wiradi, 2002. *Good Times and Bad Times in Rural Java.* Leiden: KITLV Press.

Brown, N. and M. Press, 1992. 'Logging Rainforests the Natural Way?' *New Scientist,* 14 March: 25–29.

Cleary, M. and P. Eaton, 1992. *Borneo: Change and Development.* Singapore: Oxford University Press.

Colchester, M., 1993. 'Pirates, Squatters and Poachers: The Political Economy of Dispossession of the Native People of Sarawak.' *Global Ecology and Biography Letters* 3: 158–179.

Cramb, R.A., 1988. 'The Commercialization of Iban Agriculture.' In R.A. Cramb and R.H.W. Reece (eds), *Development in Sarawak: Historical and Contemporary Perspectives.* Melbourne: Monash University, Centre for Southeast Asian Studies.

———— and I.R. Wills, 1990. 'The Role of Traditional Institutions in Rural Development: Community-Based Land Tenure and Government Land Policy in Sarawak, Malaysia.' *World Development* 18: 347–360.

Dauvergne, P., 1997. *Shadows in the Forest: Japan and the Politics of Timber in Southeast Asia.* Cambridge: MIT Press.

Douglas, I., T. Greer, K. Bidin and M. Spilsbury, 1993. 'Impacts of Rainforest Logging on River Systems and Communities in Malaysia and Kalimantan.' *Global Ecology and Biography Letters* 3: 245–252.

FAO Staff, 1957. 'Shifting Cultivation.' *Unasylva* 11: 9–11.

Fox, J., 2000. 'How Blaming "Slash and Burn" Farmers is Deforesting Mainland Southeast Asia.' *Asia Pacific Issues* 47: 1–8.

Hong, E., 1987. *Natives of Sarawak.* Penang: Institut Masyarakat.

Hurst, P., 1990. *Rainforest Politics: Ecological Destruction in South-East Asia.* London: Zed Books.

ITTO (International Tropical Timber Organisation), 1990. *The Promotion of Sustainable Forest Management: A Case Study in Sarawak, Malaysia.* Denpasar: ITTO and International Tropical Timber Council.

Jomo, K.S., 1994. 'The Continuing Pillage of Sarawak's Forests.' In *Logging Against the Natives of Sarawak.* Selangor: Institute of Social Analysis.

Kartawinata, K., S. Adisoemarto, S. Riswan and A.P. Vayda, 1981. 'The Impact of Man on a Tropical Forest in Indonesia.' *Ambio* 10: 115–119.

Kedit, P.M., 1993. *Iban Bejalai [Iban Walkabout]*. Kuching: Sarawak Literary Society.

King, V.T., 1993. '*Politik Pembangunan*: The Political Economy of Rainforest Exploitation and Development in Sarawak, East Malaysia.' *Global Ecology and Biogeography Letters* 3: 235–244.

Lau, B.T., 1979. 'The Effects of Shifting Cultivation on Sustained Yield Management for Sarawak National Forests.' *Malaysian Forester* 4: 418–422.

Majid Cooke, F., 2002. 'Vulnerability, Control and Oil Palm in Sarawak: Globalization and a New Era?' *Development and Change* 33: 189–211.

Marn, H.M. and W. Jonkers, 1982. 'Logging Damage in Tropical High Forest.' In P.B.L. Srivastava, A.M. Ahmad, K. Awang, A. Muktar, R.A. Kader, F.C. Yom and L.S. See (eds), *Tropical Forests: Source of Energy through Optimisation and Diversification*. Serdang: Universiti Pertanian Malaysia.

Mertz, O. and H. Christensen, 1997. 'Land Use and Crop Diversity in Two Iban Communities, Sarawak, Malaysia.' *Danish Journal of Geography* 97: 98–110.

Myers, N., 1992. 'Tropical Forests: The Policy Challenge.' *Environmentalist* 12: 15–27.

Ngidang, D., 2002. 'Contradictions in Land Development Schemes: The Case of Joint Ventures in Sarawak, Malaysia.' *Asia Pacific Viewpoint* 43: 157–180.

Potter, L., 1993. 'The Onslaught on the Forests in South-East Asia.' In H. Brookfield and Y. Byron (eds), *South-East Asia's Environmental Future: The Search for Sustainability*. Tokyo: United Nations University Press.

Rasul, G. and G.B. Thapa, 2003. 'Shifting Cultivation in the Mountains of South and Southeast Asia: Regional Patterns and Factors Influencing the Change.' *Land Degradation and Development* 14: 495–508.

Rigg, J., 2001. *More Than the Soil: Rural Change in Southeast Asia*. Harlow: Prentice Hall.

Ross, M.L., 2001. *Timber Booms and Institutional Breakdown in Southeast Asia*. Cambridge: Cambridge University Press.

Sidu, J., 1995. 'Socio-Economic Survey of Local Communities.' Kuching: Forest Department Sarawak.

Wadley, R.L., 1997. Circular Labor Migration and Subsistence Agriculture: A Case of the Iban in West Kalimantan, Indonesia. Tempe (AZ): Arizona State University (Ph.D. thesis).

——— and O. Mertz, 2005. 'Pepper in a Time of Crisis: Buffering Strategies, Smallholder Response and the Asian Economic Crisis.' *Agricultural Systems* 85: 289–305.

Watson, J.W., 1989. 'The Evolution of Appropriate Resource-Management Systems.' In F. Berkes (ed.), *Common Property Resources*. Belhaven: Pinter.

Whitmore, T.C., 1998. *An Introduction to Tropical Rain Forests*. Oxford: Oxford University Press.

Part IV. Conclusion

Chapter Ten

Concluding Remarks on the Future of Natural Resource Management in Borneo

Cristina Eghenter

The strength of this volume, as mentioned in the Introduction, is in its comprehensive focus on the island of Borneo (both Indonesian and Malaysian sides) as a complex and dynamic case study in natural resource management, devolution, antagonism between central state policy and community rights, and the interrelated economic and social implications at the local level.

The idea of natural resource management as a privileged 'locus' of research, analysis, and policy advocacy is certainly not new. Nevertheless, this volume contributes important perspectives and indicates, implicitly or explicitly, some key elements that should be considered by analysts and scholars, practitioners and policy makers, in efforts to promote sustainable management of natural resources in the future.

'Localised' Interventions

All the chapters reaffirm the centrality of 'locality', not only in terms of research interest (field-based 'local voices') but also as a dimension for effective political and economic reform with regard to natural resource management. The legislation granting (political and financial) regional autonomy, especially in post-Suharto Indonesia, in itself reiterates the principle that management effectiveness can be correlated positively with knowledge and recognition of local conditions, needs and solutions.

As some of the chapters of this volume in commendable ways, more attention to the local level can reveal the extreme complexity and diversity of interactions between people and natural resources: initiatives by communities to protect forest resources (Vaz, Chapter 7), viable timber-extraction schemes between loggers and swidden cultivators (Pedersen et al., Chapter 9), as well as challenges such as 'illegal' logging by communities on the border between Sarawak and West Kalimantan (Wadley, Chapter 6). Micro-level analysis allows us to determine how local factors interact with outside factors to produce particular patterns of forest use and condition. This concern with the local level can help us to deconstruct ambiguous definitions by relating the phenomenon in question ('illegal logging') to specific events and policies at various levels. It also alerts us to existing 'misperceptions' of local practices and aspirations by the central authorities to justify large development schemes (Majid Cooke, Chapter 2). The understanding generated by micro-level analysis can thus help identify appropriate solutions.

After all, it is the 'local level' (the actions of the local government and local people) that can determine the success or failure of arrangements in natural resource management. As stated by Gibson, McKean and Ostrom, 'forest management is intensely local' (2000: 21).

Intervening locally or advocating localised management models of natural resource management does not imply that solutions are only valid for a specific site or a specific group of people, nor does it imply that everything local is necessarily good management or good for conservation. 'Local' solutions have to work at a local level. For example, a 'localised' intervention would be the drafting of flexible policies that allow for the integration of local practices and knowledge — policies based on a thorough understanding of complex local dynamics and changing local conditions. 'Localised intervention' could be made possible by legal pluralism and local policies that recognise the rights and claims of traditional communities (Khan 2001; Casson, Chapter 4).

The analyses in this volume also point to the importance of the political, legislative, and economic space 'in-between', a space not strictly regulated by the state but dominated by uncertainty, an economic landscape where multiple and conflicting claims exist. This kind of space is often rich in experimentation and local initiatives. Whatever regulations might be issued by the central state, local communities might both uphold and adapt the spirit of the law, and they might filter or ignore outside regulations. They can resist persuasion (Majid Cooke and Bulan, Chapters 2 and 3). They can also generate their own regulations. They can negotiate little-used provisions in the law to their advantage (Vaz, Chapter 7). This is especially the case with *adat* and the resilience of indigenous resource-management systems that filled a void in state legislation and managed to guarantee sustainability of large tracts of forest under stable conditions (Eghenter, Chapter 8). These studies demonstrate the need to appreciate such local-level variation, and apply or adapt solutions that would work at the local level.

The transition to decentralisation, as well as increased opportunities for exploitation of natural resources, might have generated conflicting situations and a lack of transparency in power-sharing arrangements between communities, the private sector, and local and central governments. In Malaysia, hypermodernist ideologies of development have pushed the implementation of development programs based on plantation expansion and land estate schemes. This requires the use of persuasion — which could include intimidation (Majid

Cooke, Chapter 2). However, at the local level, at the point of encounter, other factors might intervene and persuasion may end up working in favour of government or, on occasion, against it, and in favour of local communities (Bulan, Chapter 3).

For all these reasons, 'optimism in locality is not misplaced', to paraphrase an expression on community by Agarwal and Gibson (1999). 'Localised' remains the appropriate dimension for reform to promote good governance and equity in natural resource management.

Relevant Research

Although none of the authors explicitly deals with the issue of the relevance of research, the question of whether their analyses might be useful, lead to good legislation, or help to defuse conflicts in natural resource management, is, I believe, an unexpressed yet strong concern of all the authors. For one thing, many of the chapters end with explicit recommendations on what ought to be done (Wadley, Vaz and Eghenter, Chapters 6, 7 and 8).

What is the role of research in identifying appropriate solutions in natural resource management? And what kind of analysis is most needed? The divide between critique and engagement — between, on the one side, scholars content with placing a critical gaze on policies and models of natural resource management, and, on the other, practitioners having to negotiate a difficult course of action to promote better natural resource management — has only sharpened recently. The divide is not only one of 'position' between those situated in the field (working for local, national or international NGOs, or for government agencies) and those based in academic institutions, but also one of methodological approach and priorities, language and definitions.

The analysis that is needed is indeed one that can break down barriers and provide a bridge between research and management. The researchers needed are those who can easily straddle critique and engagement, and who can link critique to the formulation of viable solutions, and influence engagement with detached assessment. The research should ask good questions and adopt an open and exploratory mode to produce results that help policy makers, practitioners and analysts to understand the situation and exercise 'good judgment' about possible solutions (Sayer and Campbell 2004).

It is important, however, to avoid assuming that such an exercise is simple or linear. The approach needs to consider the complexity of the situation of dynamic landscapes where different groups of actors have their stakes or claims. It might require an initial assessment of the situation to identify existing policies, relevant actors, interests and structural factors. It might also require a high degree of flexibility and a multiple-entry (or multi-hypotheses) strategy whereby several lines of inquiry may be pursued (Eghenter 2003). Policy-oriented research

State, Communities and Forests in Contemporary Borneo

requires a methodological approach that enables the formation of context-specific generalisations; that is, explanations or descriptions of the causal interactions of various factors that obtain within the specificities of a given situation.

There is still an information gap about accepted, scientific understanding of which variables are the primary causes of deforestation, loss of biodiversity, and so on (Gibson, McKean and Ostrom 2000). The purpose of research and analysis with regard to natural resource management is ultimately to try to understand complexities rather than aspire to produce complete explanations (Sayer and Campbell 2004).

Multi-stakeholderships

Tenure insecurity and vulnerability, overlapping claims and volatile situations, over-exploitation of resources and degradation, unfair distribution of benefits and other problems regarding natural resource management cannot be addressed unless all stakeholders are equally and equitably involved. This calls for more than a generic commitment to participation or an attempt to work with stakeholders (Brosius and Russell 2003). It implies building partnerships and strengthening the constituencies upon which the legitimacy of partnerships is based. Clear 'rules of engagement' should frame the roles, responsibilities and rights of each party. Often, statements of principle such as 'community forestry' or 'collaborative management' might remain empty slogans unless there is political will, agreed guidelines and mechanisms to secure fair participation among partners and avoid asymmetrical relationships of power and privilege (see Deddy, Chapter 5).

One of the most commonly upheld hypotheses in good forest management is that local forest users should participate in designing, and have authority over, the institutions that govern the use of natural resources, as well as having the right to participate in modifying these rules (Gibson, McKean and Ostrom 2000: 253). These rules, however, should be based on principles of sustainability and benefit sharing, and they should be enforced. Where rules or policies are perceived as biased, unclear, or ignored, sustainable natural-resource management cannot be guaranteed. Policy reform to address key issues such as tenure insecurity and unsustainable exploitation is necessary (Wadley, Chapter 6). The example of the West Kutai Regional Forestry Program Working Group (Casson, Chapter 4) is indicative. A consultation process and forum where key stakeholders are able to participate, discuss the complexities of natural resource management, and develop appropriate strategies could facilitate the formation of enduring constituencies. It could also foster the creation of accountable and equitable multi-stakeholderships for the sustainable management of natural resources.

200

References

Agarwal, A. and C.A. Gibson, 1999. 'Enchantment and Disenchantment: The Role of Community in Natural Resource Conservation.' *World Development* 27: 629–649.

Brosius, P. and D. Russell, 2003. 'Conservation from Above: An Anthropological Perspective on Transboundary Protected Areas and Ecoregional Planning.' *Journal of Sustainable Forestry* 17(1/2): 39–65.

Eghenter, C., 2003. 'What Kind of Anthropology for Successful Conservation Management and Development?' Paper presented at the meeting of the American Anthropological Association meeting, Chicago, 17–23 November.

Gibson, C., M. McKean and E. Ostrom (eds), 2000. *People and Forests: Communities, Institutions, and Governance*. Cambridge (MA) and London: MIT Press.

Khan, A., 2001. 'Preliminary Review of Illegal Logging in Kalimantan.' Paper presented at the conference on 'Resource Tenure, Forest Management and Conflict Resolution', Canberra, 9–11 April.

Sayer, J. and B. Campbell, 2004. *The Science of Sustainable Development: Local Livelihoods and the Global Environment*. Cambridge: Cambridge University Press.

Index

adat
  communities, 100, 170, 171
    in Kayan Mentarang National
    Park, 164–168
  criteria for natural resource
  management, 166–168
  (customary) law and practice, 36, 37,
  90, 163
    'lieutenant customary law', 93
  leaders, 11, 79, 99, 117
    land managed under (Kalimantan),
    13
  social, environmental and legal
  dimensions, 163–180
  rights, 89
Agrarian Law 1960, 82
Agreed Forest Land Use Plan, 7
agricultural policies, 51–52
alienation of land, 52, 53, 157
Alimuddin, Sultan, 75
Animal Husbandry Improvement Scheme,
  189
Application of Law Ordinance 1949, 47
Association of Indonesian Timber
  Concession Holders, 78

Ba Kelalan by-election, 30, 38
balok, 121, 127
Basic Forestry Law, 7, 91, 92, 116
Belize, 95
bicycle logging, 120–121, 127
biodiversity
  and community interests, 134
  and communal lands, 155–156
  conservation, 157, 158, 163
  loss of, 157, 200
  seeking spaces for, 133–162
Bock, Carl, 69
borderlands, 112, 127
  Upper Kapuas, 112–117, 118
'borderlanders', 11, 112–113, 115, 128
borders, 4
Borneo
  conservation interventions in, 3–21

economic activities, 4
  future of natural resource management,
  197–201
  Indonesian, 3, 4, 5
  Malaysian, 3, 4, 5
boundaries
  administrative, 102, 103
  ancestral, 102, 104
  delineation of, 106
British North Borneo Company, 142
Brooke, James, 46, 113, 114, 115
  family, the, 45
  period, the, 27, 33, 34, 41, 58

cash crops, 35, 147, 188, 189
Central Land Council (Australia), 95
Centre for International Forestry Research
  (CIFOR), 65, 99, 101
cockfight, 124, 125
coconut, 71
collaboration with government partners,
  154–155, 159, 177
colonial period, 35, 36, 41
commercial crops, 150
communal lands
  and biodiversity, 155–156
community-based management, 96, 106,
  107, 134, 148–150
community–conservation partnerships,
  134, 147–150, 157–159
community cooperatives in West
  Kalimantan, 111–132
    'directly liaised', 119–120
    'indirectly liaised', 119–120
Community Empowerment Team, 166
community forestry, 79–81, 84, 155, 200
Community Forestry Permit, 81
community mapping, 14, 15, 16, 89–110,
  124, 172
  as tool to reduce conflict, 101–102
  for protected area management, 98–99
  for recognising indigenous rights, 100
  for research objectives, 99–100
  implications, 94–106
  vested interests behind, 104–106
community participation, 14, 95

risks to, 14
conflict, 89, 150
  between communities, 103
  of interest, 97
  over land and natural resources, 91–94,
    101, 107, 111, 124, 182
  resolution, 4, 89–110
  tenurial, 95
  with logging companies, 105, 140
  with state, 89
conservation, 9, 14, 95, 151, 157, 176
  as 'neo-colonial' project, 9
  as umbrella, 14
  biodiversity, 133
  community support for, 136
  management of, 170
  on community lands, 155–159
  through land tenure security, 143–144
  under *adat* in East Kalimantan, 163–180
cooperatives, 169
'co-optation', 11, 39
corruption, 14, 111, 123
'counter-mapping', 90
customary councils, 165, 166, 174

Danau Sentarum National Park, 117, 124
Danish Agency for Cooperation and
  Development, 135
Dayak Benuaq villages, 72
Dayak groups/peoples, 26, 28, 30, 31, 36,
  37, 115, 164
  backwardness, 32, 33
  'vulnerability', 31, 32, 33
decentralisation, 3, 25, 83, 198
  and management of National Parks,
    172–173
  forests and estate crops, 65–86
  laws, 77
  oil palm, 81–83
deforestation, 66, 200
Department of Agriculture, 189
Department of Environmental
  Conservation, 156
Department of Lands and Surveys, 142,
  145, 154
depersonalisation of social life, 26

development
  alternative approaches to, 5
  and Native Customary Land, 35–38
  community, 10
  economic, 134
  large-scale activities, 90, 94
  participatory, 10
  social and political, 11
  state-driven, 5–9
  sustainable, 11
  the state and localities, 41–42
  top-down, 4, 10, 38
Directorate General of Forest Protection
  and Nature Conservation, 175, 177
District Land Use Plan, 70
donors, 78, 83
Dutch, the, 75, 91, 113, 114
Dutch–Sarawak rivalry, 116

East Kalimantan
  changes to adat tenure in, 92–94
  division of, 66
  districts of, 66
economic crisis (1997), 111
élites, 94, 106, 111, 120
English common law, 47
estate crops
  agro-industrial, 71–74
  decentralisation and forests in Kutai
    Barat, 65–86
ethnic difference, 15

Federal Land Development Authority
  (FELDA), 51
Federation of Malaysia, 116
fish ponds, 189
forest conversion, 90, 124
forest degradation, 6, 123, 133, 150, 154
forest fires, 70, 90, 94, 106
Forest Land Use Consensus Plan, 70, 91
forest products, 164, 188
  collection of, 166
  non-timber, 14, 127
forest release permits, 82
forest regeneration, 187
forest resources, 77, 81, 158

Forest Utilisation and Forest Product Harvesting in Production Forests, 77, 128
Forestry Law 1999, 170–172
forests and estate crops, 65–86
Forum Musyawarah Masyarakat Adat, 173–177

gaharu trees, 167, 168
geographic information systems (GIS), 97, 100
geomatics, 97, 98
globalisation, 32, 81
gold mining, 76, 77
Government Regulations on Conservation and Protected Areas, 168
government, role as trustee, 28
GPS, 100, 104
Habibie government, the, 91
Hak Pemungutan Hasil Hutan (HPHH) concessions, 66, 77, 78, 79, 83, 84, 94, 105
Hak Pengusahaan Hutan (HPH) concessions, 70, 71, 75, 78, 84
headhunting, 115, 137, 144
High Modernism in Sarawak, 27–31
hunting and trapping, 166
Hutan Tanaman Industri (HTI), 71
hypermodernist ideologies, 198

Iban communities /peoples, 115, 124, 182
    impact of logging on, 183
    socio-economic conditions in, 188–189
Iban concession, 117
'illegal squatters', 36, 37
illipe nut tree, 189
income-generation schemes, 15, 189
indigenous
    cultural traditions, 12
    land rights, 13
    management systems, 90, 94
    on being, 12–16
Indonesia
    management of National Parks in, 168–170
Indonesian Institute of Sciences, 172
International Tropical Timber Organisation, 183

Japanese conglomerates, 182
joint ventures, 28, 37, 190
    as new model, 52–54, 59
    economic viability of, 39–40
    questions of proof under, 60–61

Kalimantan, 5, 8
Kayan Mentarang National Park, 15, 98, 99, 100, 163, 172, 173, 174, 176, 177
    adat communities in, 164–168
kerangas forest, 135, 142
Konfrontasi, 116
Konsep Baru (New Concept), 9, 13, 26, 27, 28, 30–42
krismon (economic crisis), 117, 118, 120, 123
Kutai Barat, 11, 66, 83
    decentralisation, forests and estate crops in, 65–86
    district finances, 74–77
    forest resources in, 69–71, 175
Kutai Kartanegara, 66, 68, 83, 102
Kutai Timur, 66

labour migration, 119, 127, 189
Land (Classification) Ordinance (1948), 35, 47, 151
    zones, 47–48
Land Code 1958, 34, 35, 36, 48, 49–51
    Subsection 5(2)(f), 37, 38, 49
    1974 amendments to, 7
    1988 amendment, 52
    Amendment of 2000, 27, 28, 30, 49
Land Consolidation and Development Authority (LCDA), 52, 59
Land Consolidation and Development Authority Ordinance 1981, 52
land
    alienation of, 52
    certificates, 28, 29, 30
    community access, 15
    conflict over, 89, 106
    customary access to, 13
    customary/indigenous tenure of, 4, 29, 49, 137

ambiguities in interpretation,
    145–146
        changes to, 92–94
        complexity of, 103
    development schemes and agricultural
    policies, 51–52
    'idle', 'unoccupied', 'waste', 13, 27, 32,
    33–35, 105
    indiscriminate clearing of, 150
    individual claims, 165
    individual titles, 152
    native customary rights to, 45–51
    registration of, 28
    reserved forest (see also 'pulau'), 47
    tenure
        conflict over natural resources,
        91–94
        conservation through, 143–144
        insecurity of, 13, 29, 142, 154, 200
    virgin, 46
Land Regulations 1863, 34
Land Surveyors Ordinance 2002, 61
Land Use Delineation Law, 99
log ponds, 75
logged-over forest, 6, 190
logging, 4, 35, 69
    camps, 142, 149, 189
    coexistence with shifting cultivation,
    181–193
    concessionaires (see also 'tukei'), 12, 15
    concessions, 7, 30, 71, 116, 126
    disputes on Indonesian–Malaysian
    border, 125–126
    environmental impact, 6
    illegal, 7, 70, 93
    'illegal', 11, 126, 197
        in West Kalimantan, 111–132
        and regional autonomy, 117–126
    interests, domination of, 16
    low-mechanised, 112, 187
    Malaysian companies, 118
    selective, 186–187
longhouse communities, 39, 40, 41, 46, 124,
    141, 183, 184, 187
Lotaq community, 103, 104, 105
Lundayeh, 12, 134, 137, 144, 146, 151

urban, 145

Mahakam River, 75, 78
Malaysian Human Rights Commission, 30
Malinau River, 99, 100, 101
map making, 14
Marhum Pemarangan, 102
migration, 12
Ministry of Environment, 91
Ministry of Forestry and Estate Crops, 79,
    92, 122, 168
Ministry of Forestry and Plantations, 172,
    177
Ministry of Home Affairs, 91
Ministry of Tourism and Environmental
    Development, 135
Model Forest Management Area (MFMA),
    15, 183–185, 186, 188
modernity vs. conservatism, 31
Muara Begai community, 103

National Development Planning Agency
    (BAPPENAS), 82, 91
National Land Agency, 82
Native Customary Land (Sarawak), 3, 13,
    25, 27, 28, 30, 61
    and 'development', 35–38, 55–56
    'idle', 25–44
    registration of, 36
    security of tenure of, 39
    trusts as device for land development,
    45–64
Native Customary Rights Land (Sarawak),
    181, 183, 187
Native Reserve, 152, 153, 154
Native Title Land (Sabah), 152, 155
natural resource management, 197–201
Natural Resource Management Project, 79,
    172
    localised models of, 197–199
natural resources, new guidelines for use
    of, 169
'neo-colonialism', 10
Netherlands Indies Government, 115
NGOs, 10, 77, 83, 90, 97, 154, 155
    Indonesian, 100

international, 102, 199
local, 102, 105, 128, 129
role of, 4
'Nine Cardinal Principles of the Rule of the English Rajahs', 58

off-farm jobs, 186
oil palm sector
decentralisation and, 81–83
as economic saviour, 9
as environmental vandal, 9
'green gold', 9
plantations, 117, 126
production, 4, 8, 25, 35, 66, 71
Orang Asli, 58
otonomi daerah, 111, 126
'outside investors', 146
overseas entrepreneurs, 82

padi farming (*temuda*), 36, 50
participation, 14, 15, 200
partnership(s), 5, 200
community–conservation, 134
pepper, 15, 35, 118, 127
Permanent Forest Estate, 183, 187, 190
Permits to Use and Harvest Timber, 94
place
as a commodity, 12, 31
attachment to, 12, 13
plantation agriculture, 25, 32
plantation companies (see also 'HTI'), 31, 71
plantation estates, 76
PLASMA scheme, 73, 79
post-colonial era, 35
power relations, unequal, 11, 12, 15
power-sharing arrangements, 198
property, 89
Provincial Forestry Service, 69, 70
Provincial Land Use Plan, 70
PT Kelian Equatorial Mining (PT KEM), 76, 77, 83
PT London Sumatra International Tbk (PT LonSum), 72, 73, 82, 83
plantation area, 74
PT Sarana Trirasa Bakti, 99

PT Yamaker, 116
pulau, 25, 36, 47, 50

Rajah in Council, 47
Rama Alexander Asia, 67, 79
rattan, 93, 94, 165, 166
'realisable utopias', 27
recreational hunters, 142, 149
regional autonomy, 78, 122
in West Kalimantan, 111–132
and 'illegal' logging, 117–126
Regional Land Use Plan, 92
relevant research, 199–200
resource management
adat criteria for, 166–168
by state, 15
participatory, 14
top-down, 15

rubber, 35, 71

Sabah, 5, 6, 8
Sabah Biodiversity Conservation Project, 135, 152
Sabah Conservation Strategy, 135, 140
Sabah Forest Industries (SFI), 140, 141
Sabah Land Ordinance 1930, 141
Sarawak, 5, 6, 7
hypermodernist tradition, 36, 38
land-development policy, 3, 45–64
Sarawak Forest Department, 183
Sarawak Land Consolidation and Rehabilitation Authority (SALCRA), 51, 52
Sarawak Land Development Board (SLDB), 51, 52
shifting cultivation, 7, 34, 164, 185
coexistence with commercial logging, 181–193
negative attitudes towards, 182
siltation, 188
Special Directive on the Management of National Parks, 168
state, the
control over land, 35
institutions as agents of change, 16

management of natural resources, 15
strategies of control, 42
State Land, 142, 145, 146, 150, 190
state 'persuasion' processes, 27, 31–33, 41, 198
limits of, 38–41
state spaces
expanding, 25–44
strategies for, 26–27
Suharto
era, 90, 9
fall of, 94, 117, 123, 126
family, 7
government, 69, 71
New Order regime, 11, 14, 111, 116
Sukarno, President, 116
Sulaiman, Sultan Mohammed, 74
sustainable forest management, 112, 176, 182, 200
sustainability, 163, 197, 200
swidden farming, 46, 47, 103, 127, 137, 141, 150, 197

taukeh (see tukei)
Temporary Occupation Licence, 146
tenure systems
state-imposed, 91–92
traditional, 144–145, 157
tenurial rights
community mapping and, 89–110
and conflict resolution, 89–110, 156
territorialisation, 26
timber, 5
concessions, 89
large-scale production, 4, 6, 70
low-impact harvesting, 121, 127
smuggling of, 111, 126
state-run companies, 75
Toledo Alcaldes' Association, 95
Toledo Maya Cultural Council, 95
Torrens system, 35
tourism, 156, 157
ecological, 12
trade-offs, 31, 38–41
transmigrant settlements, 89
trust

and Native Customary Land
development, 55–56
and protection of property, 54–62
as device for land development, 45–64
breach of and remedies for, 59–60
nature of, 62
trustees
fiduciary relationship, 57–59
powers and duties of, 56–57
tukei, 12, 118, 120, 123, 125, 126

Ulu Padas, 135, 149, 150, 156, 158
logging interests in, 140
State Land, 150, 152
Ulu Padas Commercial Forest Reserve, 137, 139, 156
Ulu Teru longhouses, 39, 40, 41
United States Agency for International
Development (USAID), 79, 172

'virgin jungle', 36, 49

West Kutai Regional Forestry Program
Working Group (KKPKD), 79, 200
wet-rice farming, 127, 164
Worldwide Fund for Nature (WWF), 164
WWF Indonesia, 98, 99, 100, 172, 175
WWF Malaysia, 133, 134, 135, 141, 143, 147, 148, 151, 158

zonation system, 168, 176

www.ingramcontent.com/pod-product-compliance
Lightning Source LLC
Chambersburg PA
CBHW061240270326
41927CB00035B/3447

9 781920 942519